Precision technology and sensor applications for livestock farming and companion animals

Precision technology and sensor applications

for livestock farming and companion animals

edited by:

E. (Lenny) van Erp-van der Kooij

EAN: 9789086863648
e-EAN: 9789086869176
ISBN: 978-90-8686-364-8
eISBN: 978-90-8686-917-6
DOI: 10.3920/978-90-8686-917-6

First published, 2021

© Wageningen Academic Publishers
The Netherlands, 2021

This work is subject to copyright. All rights are reserved, whether the whole or part of the material is concerned. Nothing from this publication may be translated, reproduced, stored in a computerised system or published in any form or in any manner, including electronic, mechanical, reprographic or photographic, without prior written permission from the publisher, Wageningen Academic Publishers, P.O. Box 220, NL-6700 AE Wageningen, The Netherlands.
www.WageningenAcademic.com
copyright@WageningenAcademic.com

The individual contributions in this publication and any liabilities arising from them remain the responsibility of the authors.

The publisher is not responsible for possible damages, which could be a result of content derived from this publication.

Preface

Precision Livestock Farming and precision technologies in companion animal care are upcoming and in some cases almost mainstream. Just as we are depending more and more on data and technology, with smartphones, smartwatches, smart homes and smart cities, so is the livestock and companion animal sector. Many of our dairy cows have activity meters, cameras are being installed in pig and poultry farms, activity meters are used for horses and GPS trackers for dogs and cats. For all these technologies, context and domain knowledge are extremely important. If you do not know about the biology and management of the animals you are working with or designing technology for, I believe that your project will not succeed. At the same time, working in livestock production or companion animal care is difficult, if not impossible, without some knowledge of data and technology. There are not many study books available that combine precision technology and animal knowledge. With this book, we aim to combine these knowledge areas. We describe the main animal sectors in livestock, companion animals and horses, and we give an overview of the application of technology and sensors for these sectors. We also describe the main trends and concerns in each sector, since these are important drivers of the technology. Separate chapters focus on the working mechanism of sensors and the data science applications. The purpose of this book is to help students understand the cross-over field between technology and biology. The book is meant for students from an animal science or biology background that want to work with data and technology, but also for students from a business, technology or IT background that are venturing into the animal sector.

The writing of this book was a team effort. It was a great pleasure to work with colleagues from HAS University in Den Bosch, from Aeres University in Dronten, from the Universidad de Córdoba (ETSIAM) in Spain and Harper Adams University in the UK. Together we worked hard to produce this book, we learned a lot on the way, and we cannot wait to implement the book in our teaching modules and beyond. We hope that this book will inspire students and lecturers alike, and introduce them to, as well as win them over for, this very interesting field.

Preface

We are grateful to the students and farmers that provided photos for the book. Furthermore, we want to thank the reviewers that helped improve the text: Karl Behrendt, Marc Cox, Rick van Emous, Jeroen de Haas, Bart Houx, Judith Roelofs, Mark Rutter and Menke Steenbergen.

Finally, we wish to thank Wageningen Academic Publishers and Mascha Sappok in particular. It was a pleasure working with them and their enthusiasm and support made this book possible.

Lenny van Erp

Kesteren, April 2021

Table of contents

Preface   5
E. van Erp-van der Kooij

1. Introduction to precision technology and sensors   13
E. van Erp-van der Kooij

1.1 Precision agriculture   13
1.2 Precision livestock farming and companion animal care   13
1.3 Sensor technologies and sustainability   14
    1.3.1 Minimising environmental impact, minimising wasted inputs and maximising economic efficiency   15
    1.3.2 Maximising food safety and animal welfare   15
1.4 Performance of sensors   16
    1.4.1 Accuracy, precision, validity and reliability   16
    1.4.2 Sensitivity, specificity, positive and negative predicted value   18
1.5 Why invest in sensor technology?   19
References   22

2. Precision dairy farming   25
J. Roelofs and A. van den Pol-van Dasselaar

2.1 Background of dairy farming   26
    2.1.1 Overview of farm and animal numbers   26
    2.1.2 Life- and production cycle of a dairy cow   27
    2.1.3 Behaviour of a dairy cow   30
    2.1.4 Housing, feeding and milking   30
    2.1.5 Pasture-based dairy farming   32
    2.1.6 Main trends and concerns in management   33

2.2 Application of PLF in dairy farming	34
2.2.1 Precision dairy management	34
2.2.2 Precision reproduction management	37
2.2.3 Precision health management	42
2.2.4 Precision feeding management	50
2.2.5 Precision youngstock management	53
2.2.6 Precision grass management	55
2.3 Future trends in precision dairy farming	57
2.3.1 Future trends in modelling and algorithms	57
2.3.2 Future trends in sensor systems	58
2.3.3 Future trends in precision grass-based dairy farming	59
References	59

3. Precision beef and sheep farming   67

F. Maroto-Molina and D.C. Pérez-Marín

3.1 Background of beef cattle farming in Europe	68
3.2 Background of sheep farming in Europe	71
3.3 Application of PLF in beef cattle and sheep	75
3.3.1 Radio frequency identification	76
3.3.2 Global positioning systems	78
3.3.3 Accelerometers	80
3.3.4 Pedometers	81
3.3.5 Proximity loggers	82
3.3.6 Microphones	83
3.3.7 Thermistors	83
3.3.8 Weighing systems	84
3.3.9 Feed intake monitoring systems	85
3.3.10 Image-based monitoring systems	87
3.3.11 Pasture monitoring systems	89
3.4 Future trends in precision beef and sheep farming	95
3.4.1 Enhanced traceability	95
3.4.2 Energy supply	96
3.4.3 Robotics	96
References	97

4. Precision pig farming — 103
P. Jacobs and E. van Erp-van der Kooij

4.1 The pig production chain	104
4.1.1 The pig farm	104
4.1.2 Production structure in historical perspective	105
4.1.3 Regional location of the companies	107
4.1.4 Animal categories	110
4.1.5 Main trends in management	112
4.1.6 Main concerns in management	113
4.2 Application of Precision Livestock Farming in pig farming	115
4.2.1 Dashboards	115
4.2.2 Group or batch monitoring	116
4.2.3 Individual monitoring	118
4.2.4 Summary of PLF systems for group and individual monitoring	121
4.3 Monitoring the technical systems	123
4.3.1 Preventive maintenance and predictive maintenance	123
4.3.2 Monitoring group behaviour to detect events	124
4.4 Future trends in precision pig farming	124
4.4.1 Dashboards and data integration	124
4.4.2 Vision and machine learning	125
4.4.3 Sound	125
4.4.4 Practical applications	125
References	126

5. Sensors and techniques to monitor and improve welfare and performance in poultry chains — 131
A. Lourens, J.L.T. Heerkens and L. Star

5.1 Modern poultry chains	132
5.1.1 The broiler meat production chain	132
5.1.2 The layer egg production chain	135
5.2 Techniques to monitor welfare and performance	137
5.2.1 Poultry welfare	138
5.2.2 Production performance	141

5.3 Application of sensor techniques in poultry farming	142
5.3.1 Use your senses	142
5.3.2 Environmental control	142
5.3.3 Sensors to measure animal-based responses	143
5.3.4 Weighing scales	145
5.3.5 Electronic identification	147
5.3.6 Early warning systems	148
5.3.7 Camera's and images	149
5.3.8 Sound analysis	151
5.3.9 Vibration and pressure	152
5.4 Future trends in precision poultry farming	152
5.4.1 Robotics	152
5.4.2 Precision feeding system	155
5.4.3 Data-driven based on a combination of sensor and management data	155
References	158

6. Sensors and automated monitoring in horses   167
E.K. Visser and M. de Kort

6.1 Background of horses sector	168
6.1.1 Number of horses and horse enthusiasts	168
6.1.2 Equestrian entrepreneurs	170
6.1.3 Trends and concerns	170
6.2 Application of sensors and automated monitoring in horses	172
6.2.1 Physical measures	173
6.2.2 Physiological measures	176
6.2.3 Behavioural measures	180
6.3 Future trends in horses monitoring	183
6.3.1 Surface EMG	183
6.3.2 Smart textiles	183
6.3.3 Stable environment	183
6.3.4 Applications	184
References	184

7. Sensors and automated monitoring in companion animals   191
M. de Kort, K. Visser and E. van Erp-van der Kooij

7.1 Background of the companion animal sector	192
7.1.1 The pet sector: numbers and economic value	192
7.1.2 Trends in the pet sector	194
7.1.3 Concerns in the pet sector	196
7.2 Sensor applications in companion animals	198
7.2.1 Physiological measures	199
7.2.2 Behavioural measures	203
7.2.3 Sound	207
7.2.4 Summary of sensors used in companion animals	208
7.3 Future trends in technology for companion animals	209
References	210

8. Precision technology and sensors   219
E. van Erp-van der Kooij

8.1 Precision technology	219
8.2 Sensors	220
8.2.1 Acoustic sensors	220
8.2.2 Chemical sensors	222
8.2.3 Electrical current and conductivity	222
8.2.4 Flow	223
8.2.5 Mechanical sensors	223
8.2.6 Location	225
8.2.7 Optical, imaging, light	227
8.2.8 Cameras	228
8.2.9 Pressure	229
8.2.10 Temperature sensors	230
References	233

9. Data science applications in farms animals and companion animals 239
G. Hofstra and M.T.J. Terlien

9.1 Why use data science (data-driven decisions)	240
9.2 Data science workflow	241
9.2.1 Phases in data science workflow	242
9.2.2 Good practices	244
9.3 Overview of data science techniques including examples	244
9.3.1 Machine learning	244
9.3.2 Nearest neighbour and k-nearest neighbour classification	249
9.3.3 Decision tree classification	256
9.3.4 Simple linear regression	261
9.3.5 K-means clustering	264
9.4 Tooling	266
9.5 Future trends in data science applications	268
References	268

1. Introduction to precision technology and sensors

E. van Erp-van der Kooij

HAS University of Applied Sciences, P.O. Box 90108, 5200 MA 's Hertogenbosch, the Netherlands; l.verp@has.nl

1.1 Precision agriculture

Precision Agriculture has quickly become a well-known term. In arable farming, GPS is being used to drive straight with the tractor, and to irrigate, fertilise or spray in the right place at the right time. To do this, sensors are used on an increasing scale to make operations more efficient. A sensor can be described as a device that receives and responds to a signal or stimulus. Sensors can be extra eyes, ears and nose for the farmer, and these artificial senses provide the farmer with extra data. These data are translated into information, and with this information, the farmer can make better decisions. In precision farming, moisture sensors control irrigation, drones with cameras use vision techniques to recognise diseases in crops or predict yields. Satellites, weather stations and sensors provide data and information to optimise business operations. Greenhouses can be operated without human presence, and the best decisions are based on data (Hemming, 2020). Precision farming is using data to make decisions. Data-driven farming not only is more efficient, but it is also an opportunity for more sustainable crop production. Precision Farming is a way to handle resources sustainably and to work towards sustainable global food production: a food production using fewer resources, fewer pesticides and less water, because of more precise sowing, spraying and watering (Auernhammer, 2001; Fresco and Poppe, 2016).

1.2 Precision livestock farming and companion animal care

Precision agriculture in the livestock industry is also known as 'Precision Livestock Farming' (PLF); this term was introduced at the start of the 21st century, with the first conference on PLF held in 2003 (Werner *et al.*, 2003). As defined by professor Daniël

Berckmans (University of Leuven): PLF is the use of technology to automatically monitor livestock, their products and the farming environment in real-time, to aid farm management, by supplying the farmer with relevant information on which to base management decisions. The development and practice of PLF require interdisciplinary engineering and science including animal science, physiology, veterinary science, ethology, information and computing science, ICT, mechanical engineering, electronic engineering, and others. The purpose of this automatic monitoring is to detect deviations at an early stage and improve animal health, welfare and efficiency. The expected result is an improvement in overall production sustainability (Berckmans, 2014, 2017; Van Erp-van der Kooij, 2016).

There are many different technologies developed for monitoring farm and companion animals and their environment. For instance, sensors installed on an animal or in the nearby environment can detect a sudden change in animals' behaviour, such as in feeding, drinking, rumination, moving, vocalisation or productivity. Moreover, the physical state of the animal can be monitored with specific devices. Examples are monitoring body temperature, progesterone level in milk or rumen pH with thermal cameras, sensors in automatic milking stations or rumen boluses for dairy cows (ClearFarm, 2020), or monitoring body temperature and activity with sensors in the girth for horses or in the neck collar for dogs and cats.

In this book, many examples of sensor applications will be given for livestock and companion animals in Chapters 2-7. Each chapter focuses on a specific sector, and highlights are given at the beginning of each chapter. In Chapter 8, sensors and their working mechanism are described. This rather technical chapter can be used as a reference and be studied if needed. Chapter 9 describes the data science applications in the livestock and companion animal sector and gives examples of techniques used when analysing animal data. In this introduction, sustainability and the relation with the use of sensor technologies are explained in the following section. In Section 1.4 the reliability of sensors will be explained, and in the last section, the factors that influence the willingness to buy sensor technology will be discussed.

1.3 Sensor technologies and sustainability

Sustainability goals on farm level can be described as minimising environmental impact, minimising wasted inputs and thereby maximising economic efficiency, maximising food safety, and maximising animal welfare. PLF technologies facilitate these goals (Werkheiser, 2018). This promise of Precision Livestock Farming was spelled out during the final conference of the EU-PLF project: PLF has the potential to make farming more efficient by better use of resources and to guarantee or improve animal welfare (EU-PLF, 2016).

1.3.1 Minimising environmental impact, minimising wasted inputs and maximising economic efficiency

Precision feeding is a promising feeding technique to reduce the environmental footprint of livestock production systems. With precision feeding, animals within the group can be fed individually, with a feed composition better adapted to the individual animal. For example, a high-quality feed can be provided to animals that are growing faster and thus produce more efficiently, while the low-value feed is given to the animals that do not have that potential. This saves on feed costs. The practical application of individually precision feeding can have a great impact on livestock sustainability. In a study where pigs were fed individually with daily tailored diets, it was found that precision feeding reduced lysine intake by >25%, feeding costs by >8%, nitrogen and phosphorus excretion by almost 40% and greenhouse gases emission by 6% (Pomar and Remus, 2019).

Location systems for dairy cows help the farmer find a cow that needs to be checked on, treated or inseminated. This saves the farmer much time and energy. Integrating this system with other sensor systems that alert the farmer when an animal needs attention will greatly increase efficiency. Labour is an important resource on a farm, that should not be wasted; systems that decrease labour for the farmer make farming more sustainable.

1.3.2 Maximising food safety and animal welfare

Sensor systems can alert the farmer in an early stage when animals deviate from their expected behaviour or performance. This way, clinical disease is detected very early or even prevented. These 'early warning systems' result in fewer ill animals, or fewer animals being treated with antibiotics and less medication used, when the first patients are found quickly before the disease spreads any further. In other words, when farmers are alerted in time to deviations in health and behaviour, medical costs decrease and animal welfare is enhanced. The result is increased sustainability. The better we monitor the animals, the better we can take care of them, and the more sustainable the system (Matthews *et al.*, 2017).

Several sensors can monitor the physiology of individual animals to prevent or detect disease in an early stage. The biological principles of heat production during stress or inflammation can be used by measuring temperatures of certain body parts with a thermographic camera. Stress can be made visible because extremities (ears, tail) become colder and the eye region becomes warmer. An elevated temperature in claws, legs or udder can be made visible to detect lameness or mastitis (Nääs *et al.*, 2014). Temperature sensors are often combined with other sensors in an integrated system, such as in an ear tag or in an internal bolus for dairy cows, where temperature, activity and sometimes pH are combined (Mottram, 2016).

Finally, several measurements to monitor welfare can be automated using PLF systems. Housing conditions such as climate and light, feed and water, health and behaviour can be monitored with sensors, directly, e.g. by measuring body condition or body temperature, or indirectly, by measuring feed and water intake or climate conditions in the farm. Besides these animal-related factors, also to the human-animal relationship can be automatically measured. Although precision livestock technologies monitor several parameters relevant to animal welfare such as feeding and health, none of the systems yet provide the broad, multidimensional integration that is required to give a complete assessment of an animal's welfare. However, data from PLF sensors could potentially be integrated into automated animal welfare assessment systems (Larsen *et al.*, 2021; Randle *et al.*, 2017; Rowe *et al.*, 2019; Van Erp-van der Kooij, 2020)

1.4 Performance of sensors

1.4.1 Accuracy, precision, validity and reliability

Sensors are used to measure and quantify observations. For example, activity sensors are used to quantify behaviour, temperature sensors give information on the body temperature of an animal, and location sensors indicate where the animal is. Sensors provide the user with data and information to be used in the management of the farm and care of the animals. Therefore, reliable results of the measurements are essential. Only then will the results lead to good advice and possibly an improvement in animal production or animal care. Important features are accuracy (systematic error), precision (random error), validity and reliability. Accuracy is the degree of closeness to the true value and precision is the degree to which an instrument or process will repeat the same value. Validity is defined by how well the measuring device or measured value measures what it supposed to measure and reliability is how stable the measuring device is and how well the measured value detects changes in the actual value (Randle *et al.*, 2017; Streiner and Norman, 2006).

Accuracy is measured as the highest deviation of the sensor value from the true value. Inaccuracy can be represented directly in terms of the measured value, for example, a temperature sensor can have an inaccuracy of 0.15 °C. A measured temperature of 22.35 °C with this sensor, means that the true temperature is in the range of 22.20-22.50 °C. Inaccuracy can also be represented as a percentage of the measuring scale, for example, a thermometer measuring from -25 to 100 °C with an inaccuracy of 0.5% will have an absolute inaccuracy of 0.63 °C. A measured temperature of 22.35 °C with this thermometer means that the true temperature is in the range of 21.73-22.98 °C (Fraden, 2016).

Accuracy and precision are sometimes confused. Measurements are precise when the measured values are close to the average value of the quantity that is measured. In other words, when the variation is small and the random error is small. However, precise measurements can still be far from the truth. Accurate measures are measurements that are close to the true value of the quantity that is measured: in this case, the systematic error is small. However, those measurements can have a large variation. In the ideal situation, measurements are both accurate (close to the true value) and precise (with not much variation) (Lokhorst, 2018).

Precision and accuracy can be visualised with a diagram of a target pierced by some bullet holes; the tightness of the pattern of holes is a reflection of precision, and how close the centre of the pattern is to the target's bull's eye indicates the accuracy (Figure 1.1). In Figure 1A both random error and systematic error are large (or precision and accuracy are low); this means that the result lacks validity and reliability, therefore, is of little use. In Figure 1B the random error is large (or precision is low) and systematic error is small (or accuracy is high); this means that the result lacks reliability despite high validity. In Figure 1C the random error is small (or precision is high) and systematic error is large (or accuracy is low); this means that the result is reliable but lacks validity, therefore, is of little use. In Figure 1D both random error and systematic error are small (or precision and accuracy are high); overall this leads to a valid measurement and results that can be considered reliable (Randle *et al.*, 2017).

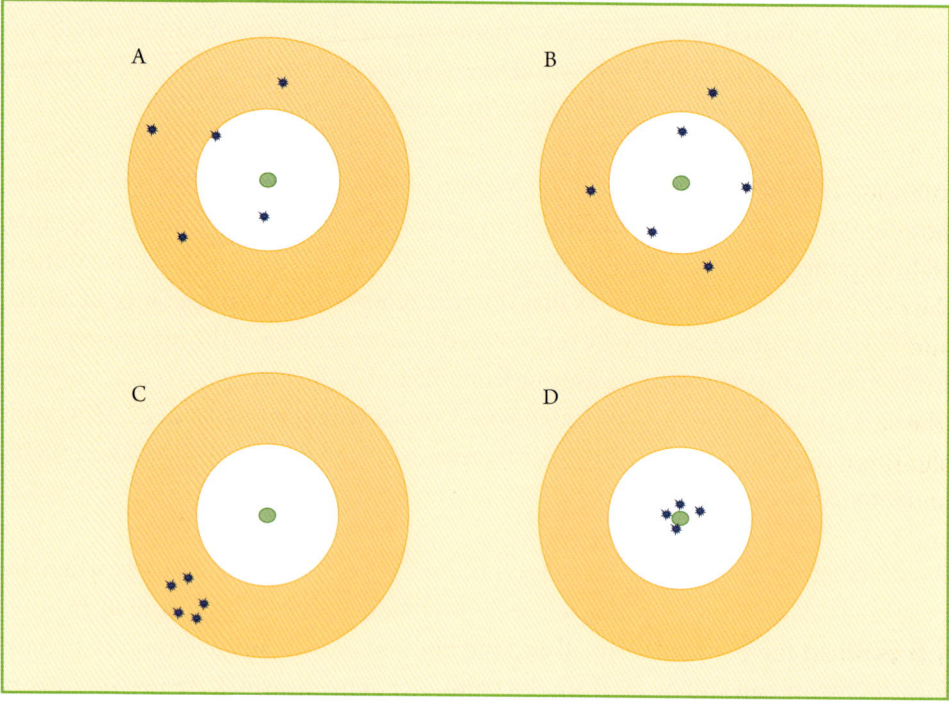

Figure 1.1. Precision and accuracy visualised, with (A) low precision, low accuracy, (B) low precision, high accuracy, (C) high precision, low accuracy and (D) high precision, high accuracy (modified after Randle *et al.*, 2017 and Streiner and Norman, 2006).

1.4.2 Sensitivity, specificity, positive and negative predicted value

When the measurements of the sensor are valid, another important question is whether the information generated by the sensor system is true or false. For example, a cow is more active during her fertile period. This increase in activity can be detected by a sensor that measures the number of steps. When the number of steps a cow takes during a certain period is higher than normal, the sensor system generates an attention, alerting the farmer the cow can be inseminated (Roelofs, 2005). While the sensor measures the number of steps with very high precision and accuracy, possibly the cow is not in her fertile period because the increased number of steps is the result of something. For example, when the claws of a cow need to be trimmed and because of that she is chased through the barn, the number of steps has increased. This increase in the number of steps is not because she is in her fertile period, but because she is chased through the barn. How well the attentions that are generated by the sensor system reflect the truth can be indicated by the sensitivity, specificity, positive and negative predictive value (Roelofs *et al.*, 2010).

When an attention is generated by the sensor system it can be either a correct attention (true positive = TP) or an incorrect attention (false positive = FP). When no attention is generated, this can be either correct (true negative = TN) or incorrect (false negative = FN). In Figure 1.2 the relation between the different attentions and the calculation of sensitivity, specificity, positive- and negative predictive value (PPV and NPV) is represented.

The sensitivity, specificity, positive- and negative predictive value depends on the algorithms that are used to translate the measures in attentions as well as on animal- and/or environmental factors that influence the variable that is measured (Roelofs and Van Erp van der Kooij, 2015).

1.5 Why invest in sensor technology?

For farmers and animal owners, the decision to invest in sensor technology depends on several factors. The main factors are economic benefits and social aspects. In a study involving questionnaires and workshops with European dairy, pig and poultry farmers, social factors and economic indicators that influenced the willingness to invest in PLF systems were determined.

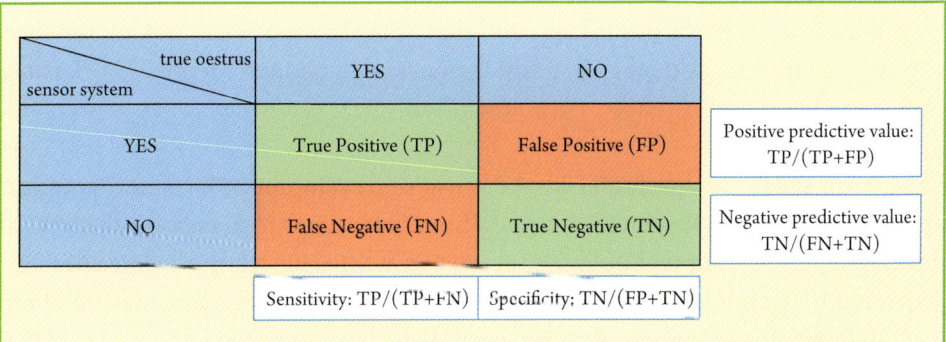

Figure 1.2. Sensitivity, specificity, positive- and negative predictive value calculation based on true positive, false positive, false negative and true negative attentions (modified from Roelofs *et al.*, 2010).

The five most important economic key indicators of PLF were feed conversion, growth, health costs, delivery weight and energy costs (Lokhorst *et al.*, 2012). This means that farmers were willing to buy PLF systems that would lower feed conversion, increase growth rates, decrease health costs, optimise delivery weight or lower energy costs. An important economic factor in the livestock sector is the profitability of a sensor system. For most devices, this has not been calculated and it is difficult to do so because of the many factors involved. The increase in the economic value of the improvement that is caused by a sensor system (i.e. the improvement in oestrus detection, milk yield, or health) determines the amount that can be invested in that sensor system (Rutten *et al.*, 2013).

The five most important social key indicators were labour conditions, number of labour hours, work pride, availability of advisory systems and the presence or absence of a successor to continue the farm (Lokhorst *et al.*, 2012). This means that farmers were more inclined to invest in technology when it would improve their labour conditions, lower their work hours, or increase the pride in their work; they were also more willing to buy systems when advisors were available to help with the new technology and when there was a successor to take over the farm in the future. Other important factors were job satisfaction, participation in a farmers' study group and social recognition for a job well done; if systems increase the job satisfaction and the farmer's work is being valued in society they are more willing to buy technology, as well as when they are part of a study group (Lokhorst *et al.*, 2012). Pig farmers in the EU PLF project mentioned that they were interested in PLF and sensor technology because they were looking for opportunities to improve their production and were interested in modern technology. Broilers farmers were interested in monitoring technologies and 'forward-thinking', and dairy cow farmers were looking for ways to improve reproduction efficiency and lameness detection in their cows (Hartung *et al.*, 2017).

Where are sensor technologies in the sense of development level and practical application? In 2013, Rutten *et al.* wrote a literature review to determine the level of development of sensor technologies for dairy cows. In this study, he describes four stages of PLF technology: level I being sensor technique, level II data interpretation, level III integration of information and level IV decision making (Figure 1.3). At that time, most sensor systems were at level I or II, which means that the system or device consisted of one or more sensors gathering data and that the data were interpreted. For dairy cows, 75% of the sensor systems under review included data interpretation, with build-in algorithms. Integration of sensor information with other information

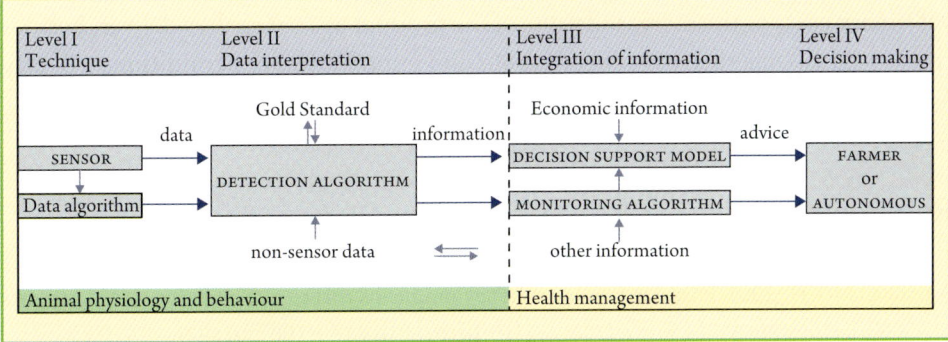

Figure 1.3. Levels of precision lifestock farming technologies, from data to information to advice. Modified from Rutten *et al.* (2013).

rarely happened (level III). In the following years, not much has changed, although there are many ongoing studies where data integration is attempted and prediction models based on aggregated data are being tested. Some systems do help in the process of decision making (level IV), such as oestrus detection systems for dairy cows (Roelofs and Van Erp-van der Kooij, 2015).

The expectations of a farmer or animal owner concerning the level of the PLF technology will also influence the willingness to invest in this new technology. If the farmer expects advice (level IV) but only gets an alert for a deviating sensor reading (level I), he will not be satisfied with his investment.

For horse and pet owners, the bond with the animal and the willingness to pay for its health and happiness drive the market. Following the trend of 'life-logging' of people, horse and pet owners are motivated to observe the activities and behaviours of their animals, to improve their wellbeing, and to understand how they act in their absence (Thompson, 2018). As we will see in Chapters 6 and 7, the market for wearable sensors for monitoring the health and welfare of horses, cats and dogs is big and expanding. Private animal owners usually do not make a conscious assessment of costs and benefits when they consider buying a device. Factors such as price, applicability, availability and brand name will play a role, but also the love for the animal and the interest in 'gadgets' and modern technology. This is somewhat different for larger, commercial companies and enterprises where more animals are housed and economic aspects are more important. For those organisations, the same factors will influence the willingness to invest in technologies as mentioned for livestock farms: economic aspects will be important, such as the possible decrease in veterinary costs or the increased value of the product or service they deliver; besides that, social aspects such as reduced labour and improved labour conditions will play a role, just as interest in modern technology.

References

Auernhammer, H., 2001. Precision farming – the environmental challenge. Computers and Electronics in Agriculture 30: 31-43. https://doi.org/10.1016/S0168-1699(00)00153-8

Berckmans, D., 2014. Precision livestock farming technologies for welfare management in intensive livestock systems. OIE Revue Scientifique et Technique 33: 189-196. https://doi.org/10.20506/rst.33.1.2273

Berckmans, D., 2017. General introduction to precision livestock farming. Animal Frontiers 7: 6-11. https://doi.org/10.2527/af.2017.0102

EU-PLF, 2016. EU-PLF project (Bright farm by precision livestock farming) – closing conference. Final Report, 233 pp. Available at: https://tinyurl.com/8pauwz7s.

Fraden, J., 2016. Handbook of modern sensors, 5th ed. Springer International Publishing AG, Basel, Switzerland.

Fresco, L.O., Poppe, K.J., 2016. Towards a common agricultural and food policy. Wageningen University & Research, Wageningen, the Netherlands, 62 pp. Available at: https://edepot.wur.nl/390280.

Hartung, J., Banhazi, T., Vranken, E., Guarino, M., 2017. European farmers' experiences with precision livestock farming systems. Animal Frontiers 7: 38-44. https://doi.org/10.2527/af.2017.0107

Hemming, S., 2020. Autonomous greenhouses. Wageningen University, Wageningen, the Netherlands. Available at: https://www.wur.nl/en/project/autonomous-greenhouses-2nd-edition.htm.

Larsen, M.L.V, Wang, M., Norton, T., 2021. Information technologies for welfare monitoring in pigs and their relation to Welfare Quality®. Sustainability 13: 692. https://doi.org/10.3390/su13020692

Lokhorst, K., 2018. Smart dairy farming. Van Hall Larenstein, Leeuwarden, the Netherlands. https://doi.org/10.31715/20181

Lokhorst, K., Vermeij, I., Bruggeman, G., Lehr, H., Hogeveen, H., Steeneveld, W., Exadaktylos, V., Gregersen, O., 2012. EU-PLF Project Deliverable 4.1 – List of socio-economic measures related to selected key indicators on farm. Available at: http://www.eu-plf.eu/wp-content/uploads/Deliverable4_1_final.pdf.

Matthews, S.G., Miller, A.L., Clapp, J., Plötz, T., Kyriazakis, I., 2017. Early detection of health and welfare compromises through automated detection of behavioural changes in pigs. Veterinary Journal 217: 43-51. https://doi.org/10.1016/j.tvjl.2016.09.005

Mottram, T., 2016. Animal board invited review: precision livestock farming for dairy cows with a focus on oestrus detection. Animal 10: 1575-1584. https://doi.org/10.1017/S1751731115002517

Nääs, I.A., Garcia, R.G., Caldara, F.R., 2014. Infrared thermal image for assessing animal health and welfare. Journal of Animal Behaviour and Biometeorology 2: 66-72. https://doi.org/10.14269/2318-1265/jabb.v2n3p66-72

Pomar, C., Remus, A., 2019. Precision pig feeding: a breakthrough toward sustainability. Animal Frontiers 9: 52-59. https://doi.org/10.1093/af/vfz006

Randle, H., Steenbergen, M., Roberts, K., Hemmings, A., 2017. The use of the technology in equitation science: A panacea or abductive science? Applied Animal Behaviour Science 190: 57-73. https://doi.org/10.1016/j.applanim.2017.02.017

Roelofs, J.B., 2005. When to inseminate the cow? – insemination, ovulation and fertilization in dairy cattle. Wageningen University, Wageningen, the Netherlands, 152 pp. Available at: https://edepot.wur.nl/28735.

Roelofs, J.B., Van Erp van der Kooij, E., 2015. Estrus detection tools and their applicability in cattle : recent and perspectival situation. Animal Reproduction 12: 498-504.

Roelofs, J., López-Gatius, F., Hunter, R.H.F., Van Eerdenburg, F.J.C.M., Hanzen, C., 2010. When is a cow in estrus? Clinical and practical aspects. Theriogenology 74: 327-344. https://doi.org/10.1016/j.theriogenology.2010.02.016

Rowe, E., Dawkins, M.S., Gebhardt-Henrich, S.G., 2019. A systematic review of precision livestock farming in the poultry sector: Is technology focussed on improving bird welfare? Animals 9: 614. https://doi.org/10.3390/ani9090614

Rutten, C., Velthuis, A., Steeneveld, W., Hogeveen, H., 2013. Invited review: sensors to support health management on dairy farms. Journal of Dairy Science 96: 1928-1952. https://doi.org/10.3168/jds.2012-6107

Streiner, D.L., Norman, G.R., 2006. 'Precision' and 'accuracy': two terms that are neither. Journal of Clinical Epidemiology 59: 327-330. https://doi.org/10.1016/j.jclinepi.2005.09.005

Thompson, R.J., 2018. The use of wearable sensors for animal behaviour assessment. PhD thesis Newcastle University, UK. Available at: https://tinyurl.com/fj8xfxaa.

Van Erp-Van der Kooij, E., 2016. Precision livestock farming – data and technology in farm animals. HAS University of Applied Sciences, Den Bosch, the Netherlands, 38 pp. Available at: https://tinyurl.com/xk4n6pjn.

Van Erp-Van der Kooij, E., 2020. Using precision farming to improve animal welfare. CAB Reviews 15. https://doi.org/10.1079/pavsnnr202015051

Werkheiser, I., 2018. Precision livestock farming and farmers' duties to livestock. Journal of Agricultural and Environmental Ethics 31: 181-195. https://doi.org/10.1007/s10806-018-9720-0

Werner, A., Jarfe, A., Stafford, J.V., Cox, S.W.R., Sidney, W. (eds.), 2003. Programme book of the joint conference of ECPA-ECPLF: 1st European Conference on Precision Livestock Farming and 4th European Conference on Precision Agriculture. Wageningen Academic Publishers, Wageningen, the Netherlands, 848 pp.

2. Precision dairy farming

J. Roelofs[1*] and A. van den Pol-van Dasselaar[2]

[1]HAS University of Applied Sciences, P.O. Box 90108, 5200 MA 's Hertogenbosch, the Netherlands; j.roelofs@has.nl
[2]AERES University of Applied Sciences, De Drieslag 4, 8251 JZ Dronten, the Netherlands

Highlights

- In Europe, the number of dairy farms has decreased, while the number of cows per farm and the milk production per cow has increased.

- Deviations from the normal behavioural pattern of a dairy cow can indicate that the farmer needs to pay extra attention to that cow.

- Since the 1980s, with the rise of automated milking systems, several devices and technologies that autonomously measure behavioural or physiological parameters of individual cows have been developed.

- Sensors are placed on or in the cow, in the close environment of the cow or in the milking machine to measure milk parameters.

- Sensors can aid in reproduction-, health-, feeding-, youngstock- and grass management.

- Systems with complementary sensors and good decision support applications are needed, which will help to overcome the limitations of single sensors and to provide better information about animals and grasslands.

2.1 Background of dairy farming

The dairy industry is constantly evolving. To get a better understanding of the sense and usefulness of sensor technology, it is important to know how dairy farming works. In the following sections, the structure and various farm processes are described.

2.1.1 Overview of farm and animal numbers

The total number of dairy cows in Europe (EU-28) was 22.6 million head in 2019 (AHDB, 2021). While the total number of dairy cows has been quite stable during the last decade (Figure 2.1), the number of dairy farms has decreased.

This means that the number of cows per farm has increased. For example, in the Netherlands, the number of farms in 2010 was almost 20,000 and decreased by 4,000 by the year 2020 (CBS, 2021). The average number of cows per farm increased from 75 in 2010 to 101 in 2020 (CBS, 2021). In the UK a decrease of 3,000 farms and an increase of 30 cows per farm was seen in the last decade (Figure 2.2, AHDB, 2021).

Not only the farm size has increased in the last decade. The milk production per cow has also increased (FAO, 2021). For example, the average milk production per cow

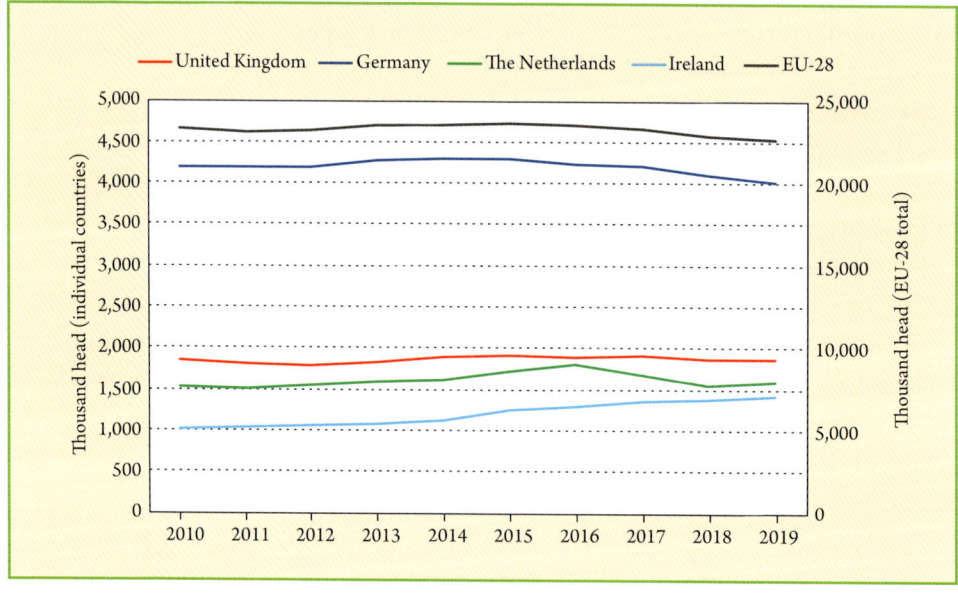

Figure 2.1. The total number of dairy cows in different countries and Europe (EU-28) (AHDB, 2021).

2. Precision dairy farming

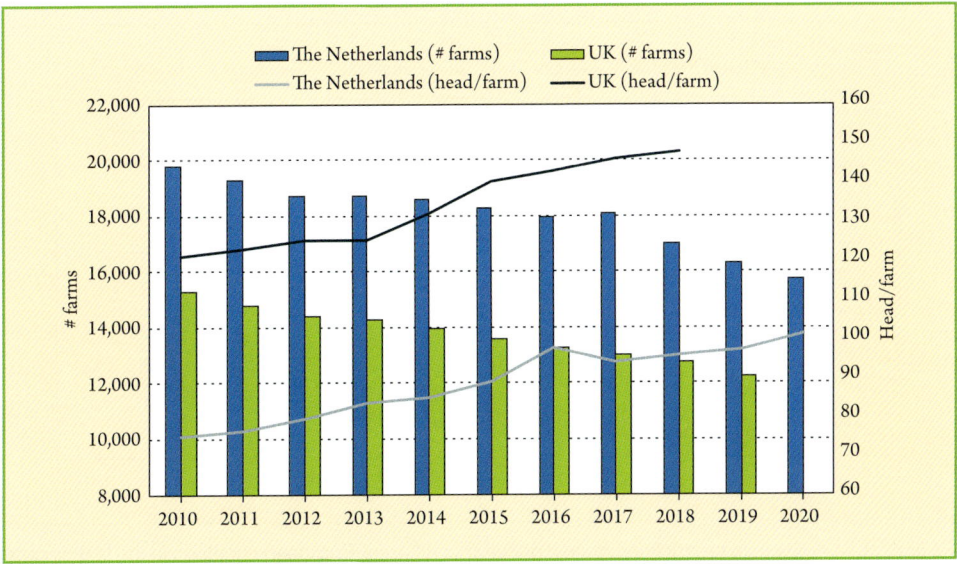

Figure 2.2. The average number of dairy cows per farm and number of farms in the Netherlands (CBS, 2021) and the United Kingdom (UK) (AHDB, 2021).

per year in the Netherlands was around 7,500 kg in 2010 and increased with 1,500 kg to over 9,000 kg in 2019. This trend is observed in many countries in Europe (Figure 2.3, FAO, 2021). Ireland is the exception in Figure 2.3. The production system in Ireland is less intensive than the production systems in other countries. Ireland has an extensive mainly grass-based system, with low input costs whereas the other countries have generally a more intensive system in which cows are fed more concentrates (Van den Pol-van Dasselaar *et al.*, 2020). Although the systems are different, which results in a different milk production per cow, the increase in production per cow per year is present in both systems. This increase in milk production comes mainly from improvements in genetics and feed efficiency (Van de Haar *et al.*, 2016).

2.1.2 Life- and production cycle of a dairy cow

A dairy cow needs to produce milk and a cow must calve to produce milk. For an optimal lactation, a cow needs to be in good health. The farmer has to provide for the needs of the cow, so she can reproduce, stay healthy and produce milk. A healthy environment with enough light and fresh air, being able to perform natural behaviour like eating, drinking, ruminating, resting and socialising and a ration suited for her lactation stage are important factors.

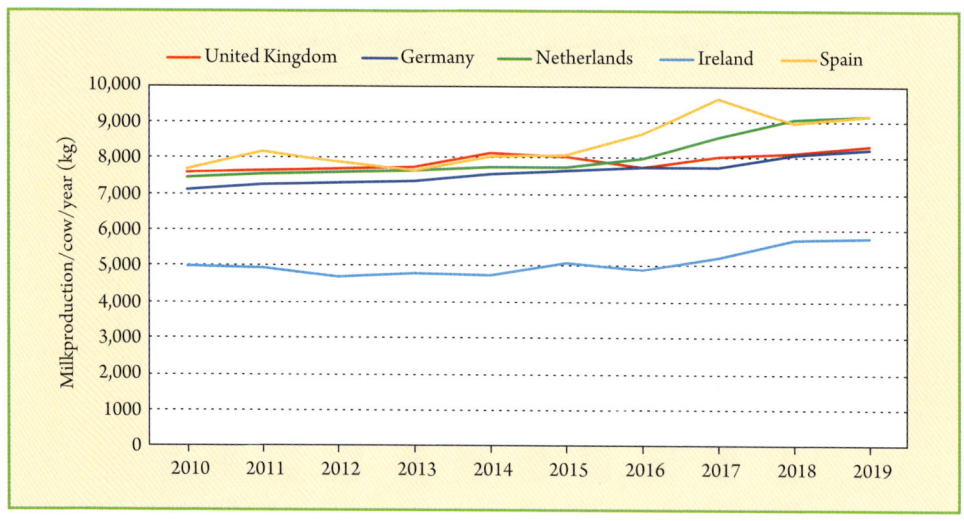

Figure 2.3. Average milk production per cow per year in different European countries in the last decade (FAO, 2021).

The foundation for a good dairy cow is laid from birth and the rearing of the young calf. A timeline with important events from birth to first calving is shown in Figure 2.4.

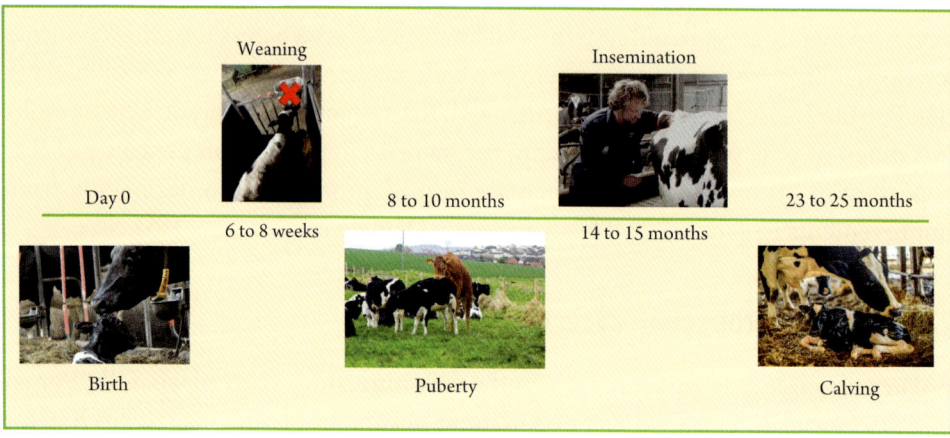

Figure 2.4. Timeline from birth to first time calving (photos: private photos and HAS University of Applied Sciences).

The first lactation starts after the first calving. The milk production increases during the first 2 months after calving and declines gradually thereafter. To maintain milk production a cow needs to calve about every year. The time interval between two calving events is called the calving interval.

The calving interval is on average 408 days in the Netherlands (CRV, 2021) and can be divided into four stages; early lactation (from calving until 3 to 4 months after calving), mid-lactation (from 3-4 months to 7-8 month after calving), late lactation (7-8 months after calving until dry period) and the dry period (6 to 8 weeks before next calving, see Figure 2.5, Strucken *et al.*, 2015).

Early lactation is a challenging period for a dairy cow. The energy needed for milk production and maintenance exceeds the amount of energy she can consume. Therefore, she needs to use her body reserves to meet the energy demand for milk production. During this stage, she also needs to recover from calving and she needs to get pregnant again. Mid-lactation is characterised by a milk production that is gradually declining. The cow is pregnant and energy demand and intake are balanced. During late lactation, the milk production declines further and body reserves can be built up. The dry period starts when the cow is dried off, i.e. lactation stops (usually 6 to 8 weeks before calving). During this period a cow does not produce milk and the mammary tissue can recover and repair before the next lactation starts.

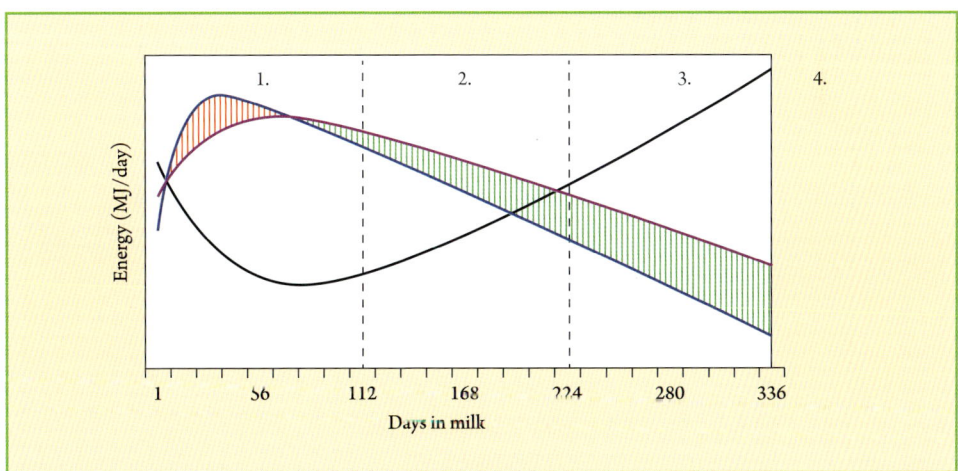

Figure 2.5. Effect of days in milk (DIM) on feed energy (purple line), energy for milk production and maintenance (blue line) and body energy stores (black line). 1: early lactation, 2: mid-lactation, 3: late lactation and 4: dry period (adapted from Strucken *et al.*, 2015).

2.1.3 Behaviour of a dairy cow

Stability and regularity contribute to the success of a dairy farm. Cows are creatures of habit and a healthy cow spends about the same time each day performing certain activities, such as eating, ruminating, lying down and walking (Table 2.1). Other activities are for example standing, grooming, agonistic behaviour, idling, drinking and being milked. Deviation from her normal behavioural pattern can indicate that the farmer needs to pay attention to that cow.

Deviation in the behavioural pattern can be an (early) warning for health problems, like foot problems, metabolic problems, udder problems or other health disorders. However, not all deviations in the behavioural pattern indicate a problem. Because of changes in reproduction hormones a cows' behaviour changes during oestrus. Oestrus is the period a cow is fertile and insemination should take place.

2.1.4 Housing, feeding and milking

Although pasture generally allows for excellent cow comfort, winter weather conditions have created the need for confinement housing in parts of the world. In 1850 the first tie-stall was patented, in 1950 bedded-pack barns were recognised as a loose-housing facility and the first free-stall barn was built in Washington State in 1961 (Bewley *et al.*, 2017). So, housing options for lactating dairy cows consist of conventional bedded-pack, tie-stall, free-stall, and compost bedded-pack barns (Figure 2.6).

Table 2.1. The normal behavioural pattern for a dairy cow kept in a free stall with cubicles and automatic milking (data derived from the practical training farm of HAS University of Applied Sciences).

Activity	Time spend on activity (h/day)
Eating	5 to 6
Ruminating[1]	9 to 10
Lying down	12 to 13
Steps	2,500-3,500 (number/day)
Lying bouts	9-11 (number/day)

[1] A cow ruminates mostly while lying down, so the hours spent lying down and ruminating are overlapping.

2. Precision dairy farming

Figure 2.6. Different housing options for dairy cattle. (A) Conventional bedded pack (photo: Shutterstock); (B) tie-stall (photo: Shutterstock); (C) free-stall (photo: courtesy of Hoeve Boveneind); (D) compost bedded-pack (photo: HAS University of Applied Sciences).

Bedded-pack and free-stall barns allow the cows to move around freely. In tie-stall barns the movement of the dairy cows is limited. Behavioural patterns of cows kept in tie-stall will differ from cows kept in loose housing systems. For example, agonistic behaviour and social interaction are less and cows spent more time standing instead of walking in tie stalls (Galama *et al.*, 2020). Nowadays most dairy cows in Europe are kept in free-stall barns. A small portion is kept in tie-stalls or bedded-pack barns (Bewley *et al.*, 2017, Galama *et al.*, 2020).

Feeding is an important factor in the production of milk. Many different feeding systems and strategies are used on dairy farms. In north-western European countries, farms mostly have feeding systems adapted for large-scale, high-yielding dairy cows that are concentrated in confinement production systems (at least seasonally). Roughage represents a major part of the feed consumed by the dairy cow. Concentrates are supplements to the roughage part of the cow's diet and provide energy and protein (typically from grains or oilseeds). Raw materials and processed (compound) feed may be used as concentrates. Roughage and concentrates can be fed separately, partly mixed or totally mixed (FAO *et al.*, 2014).

In tie-stalls cows are typically milked in their stall, the cows do not need to move. In free-stall and bedded-pack barns, cows are milked in a milking parlour or automatically by a milking robot. Conventional milking in a milking parlour requires moving the cows for each milking (usually two or three times per day) to a holding pen. The first automatic milking system was installed in the Netherlands in 1992 (Bewley *et al.*, 2017) and this system requires cows to voluntarily come to be milked. At the end of 2020 30% of the dairy farms were equipped with an automatic milking system in the Netherlands (Stichting Kwaliteitszorg Onderhoud Melkinstallaties, 2021).

2.1.5 Pasture-based dairy farming

Grasslands are primarily grazed directly by ruminants, cut for conservation or cut and carried to ruminants. Even though grazing is the traditional way of feeding ruminants, there is a clear trend of reduced grazing of dairy cows in Europe (Van den Pol-van Dasselaar *et al.*, 2020, Figure 2.7). Intensification is a major driver of this trend. In response to farm internal and external driving forces, dairy farmers choose to intensify their production systems (Oenema *et al.*, 2014), leading to more

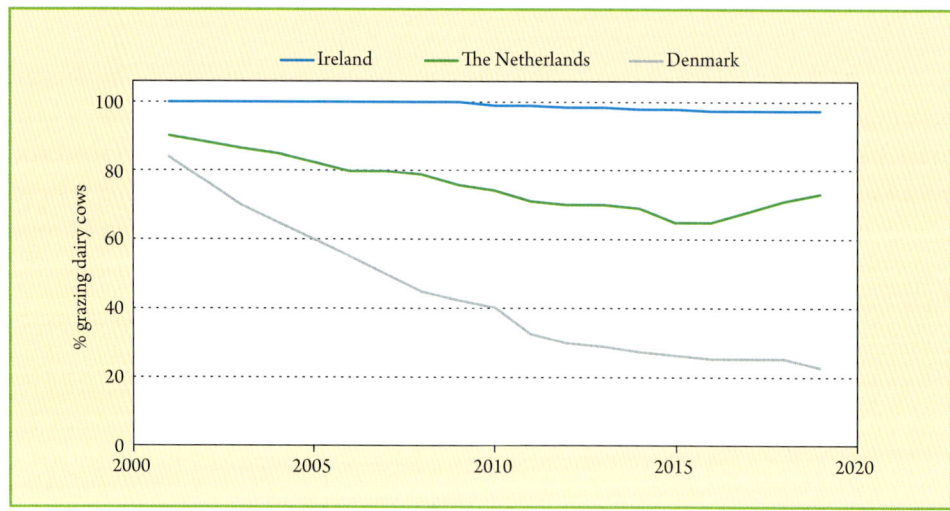

Figure 2.7. Percentage dairy cows that graze for part of the year in Ireland, the Netherlands and Denmark (educated guesses members Working Group 'Grazing' of the European Grassland Federation and for the Netherlands: CBS, 2020) (Van den Pol-van Dasselaar *et al.*, 2020).

concentrates and maize in the rations of the cows, less grass in the ration and/or less grazing. Oenema *et al.* (2014) showed that the optimum level of intensification is a moving target, due to the changing biophysical, economic and societal environments. In the Netherlands, recently the trend of less grazing of dairy cows has been changed into more grazing due to several initiatives related to a Treaty on Grazing (Runhaar *et al.*, 2020), e.g. introduction of a grazing premium by the dairy industry and enhanced commitment to grazing advice and education.

2.1.6 Main trends and concerns in management

Resilience and sustainability are keywords for the future of the dairy sector. This can be achieved with innovation, as a way to reconcile:

- the need for farmers to earn a decent living;
- consumer demand for affordable and quality dairy products;
- environmental as well as animal health and welfare requirements.

From an economic point of view, it is necessary to lower production costs to improve competitiveness. Because of price volatility and market uncertainty economic resilience has to increase for dairy farms. Simultaneously, the dairy sector must be more efficient in the use of natural resources, do more to control the environmental impact of dairy farming, take good care of herds and animals and meet health and welfare requirements (Augère-Granier, 2018).

As described, the average number of cows per farm has increased over the recent years and the number of dairy farms has decreased in many European countries. Because of better facilities and automation, one person can manage more and more cows, resulting in less time spent per individual cow. While herd size does not have a consistent, predictable association with health or welfare (Barkema *et al.*, 2015), the growing size of herds raises concerns for cow welfare by consumers and the media. In a large herd, it is important that the animals that need it receive attention. Sensors can play an important role in alerting the farmer which cows require attention and also in robust and resilient production systems.

A major challenge for pasture-based dairy farming is to meet multiple aims in a natural environment, e.g. producing food, adapting to climate change and providing additional ecosystem services like carbon sequestration and biodiversity (Delaby *et al.*, 2020). These challenges are related to the main characteristic of these farming systems, i.e. the dependency on grass as feed for dairy cows and the variability in the grass allowance, both quantitatively and qualitatively. Sensors provide opportunities to improve grazing management and pasture utilisation (Wilkinson *et al.*, 2020).

2.2 Application of PLF in dairy farming

As mentioned in the introduction of this book, PLF can be defined as 'the use of technology to automatically monitor livestock, their products and the farming environment in real-time, to aid farm management, through supplying the farmer with relevant information on which to base management decisions'. In Section 2.2.1, mainly sensors that monitor the individual cow will be discussed. In Section 2.2.2 sensors that can be used for monitoring cows' grazing behaviour on pasture and sensors that can aid in pasture management are discussed.

2.2.1 Precision dairy management

In dairy farming, the development of automatic milking systems (AMS) might be regarded as the harbinger of Precision Livestock Farming (Knight, 2020). AMS are automated systems that fitted with the trend of increasing herd sizes and the desire for less labour. Without electronic identification (EID) of individual animals, AMS and other PLF systems could not have been developed. Therefore, EID can be considered as the 'backbone' of PLF systems. Since the 1980s, work has been done on devices and technologies that autonomously measure behavioural or physiological parameters of individual cows. These sensors were developed at first to aid the farmer in deciding which cows could be bred and later development of sensors focussed and still focusses on the detection of health issues.

For a sensor to be useful for a farmer, measuring behaviour or physiological parameters is not enough. The measurements need to be interpreted, information needs to be integrated and advice has to be generated (Rutten *et al.*, 2013). Sensors enable automated, on-farm detection of changes in relevant parameters that can be related to the reproduction or health status of the dairy cow and help the farmer to perform the right actions at the right time on the right cow. Apart from collecting useful data, general farm monitoring can also be made easier and more reliable by using sensors integrated with cell phones or handheld devices instead of conventional methods, such as writing notes, keeping a farm diary, or use simple equipment without a data-sharing function. The final data can easily be viewed on a custom dashboard or office computer, which makes the technology very convenient for farmers (Neethirajan, 2017).

Nowadays many sensors are commercially available for dairy cows. Table 2.2 gives an overview of the commercially available sensors (adapted from Van Erp-van der Kooij and Rutter, 2020). The numerous sensors can be classified in different ways. Knight (2020) uses the classification 'at cow' (i.e. sensors placed on or in the cow), 'near cow' (sensors placed in the environment of the cow) and 'from cow' (i.e. sensors that collect and analyse data relating to products that come from the cow, such as milk, faeces, breath, blood etc.).

Table 2.2. Examples of precision lifestock farming technology commercially available for on-farm use (adapted from Van Erp-van der Kooij and Rutter, 2020 and 4D4F, 2019).[1]

PLF technology	Device	Recorded traits	Alerts	Company
Accelerometer	ear-tag	activity, rumination, eating	oestrus, health	Allflex, CowManager, Smartbow
	neck-tag	activity, rumination, eating	oestrus, health	Afimilk, Allflex, Cowlar Dairymaster, HerdInsights, iBO, Nedap
			oestrus, health, calving	Connecterra, Medria
		activity of a bull	oestrus	Moocall
	leg-tag	activity, lying, walking	oestrus, health, calving, wellbeing	Afimilk, ENGSsystems
			oestrus, health	Allflex, Fullwood, Icerobotics, Nedap
	reticolurumen bolus	activity	oestrus, health	Moonsyst, SmaXtec
	tail-tag	activity	calving	Moocall
Milk analysis and/or characteristics	on-line	progesterone (also in-line)	oestrus, pregnancy, reproduction	DeLaval
		LDH	acute mastitis	
		BHB	ketosis, metabolic disorders	
		SCC	mastitis	Mastiline
	in-line	conductivity, milk flow, colour, SCC	mastitis	Allflex, Boumatic, DeLaval, Fullwood, GEA, Lely, LIC Automation™
		milk solids, conductivity, lactose	mastitis	Afimilk

>>>

Table 2.2. Continued.

PLF technology	Device	Recorded traits	Alerts	Company
Temperature and/or pH	reticulurumen bolus	temperature and pH	health, calving	SmaXtec
			health	ECow, Moonsyst, Moow
		temperature	health	AgriSmart, Medria, Smartstock
	ear-tag	temperature	health	CowManager, TekVet
	neck-tag	temperature	health	Cowlar
	vaginal thermometer	temperature, expulsion	calving	Medria
Optical	(3D) camera	body condition	health	Delaval
		feeding, behaviour	health	Cainthus
		behaviour of calves	health	Futurofarming
	infrared camera	heat	mastitis	Agricam
Pressure	walk-over	ground reaction forces	lameness	Boumatic
Location	neck-tag	location in barn	no alerts, real-time cow location	Nedap, iBO
	ear-tag	location in barn	no alerts, real-time cow location	Smartbow
Scale	roughage intake control	feed intake, water intake, bodyweight, feeding behaviour	no alerts, for research	Hokofarm Group
Calf feeder		drinking and feeding behaviour	health	Biotic Industries, Delaval, Förster Technik, GEA, Holm & Laue, Lely, Urban

[1] BHB = beta hydroxybutyrate; LDH = lactate dehydrogenase; SCC = somatic cell count.

Another way to classify the different sensors is by what they measure (e.g. movement, orientation, temperature, hormones concentration). At the beginning of sensor development in dairy farming, the device often measured only one distinct feature, for example, number of steps, jaw movement or temperature (Frost *et al.*, 1997). Currently, many devices measure multiple features making classification based on what a sensor measures less feasible for the contemporary sensors. A third way to classify the sensors is by the actions or management decisions that can be made by the farmer based on the alert or advice that is generated (reproduction, health, feeding, etc.). In the following sections, this last classification method is used to discuss the different sensors.

2.2.2 Precision reproduction management

Oestrus detection and insemination time

Good reproduction is one of the keys to successful dairy farming. As already mentioned, to maintain milk production a cow needs to calve about every year. The gestation length of a cow is 280 days. This means that a cow needs to get pregnant around 80 to 100 days after calving. The main events in reproduction management are outlined in Figure 2.8.

For a period after calving the cow has no reproductive cycle, hormonal changes have to take place for the cycle to start again. Normally the first ovulation is around 3 to 4 weeks after calving (Tanaka *et al.*, 2008). A healthy cow will have an ovulation every 3 weeks when she is not pregnant. Eight to twelve weeks after calving the cow will be inseminated for the first time. Many dairy cows in north-western Europe are not inseminated by a bull but are artificially inseminated. When artificial insemination is used, the farmer needs to detect when the cow is fertile, i.e. when she is in oestrus. Not all cows get pregnant from the first insemination. Detection of oestrus or pregnancy diagnosis after insemination is important to see whether a cow is pregnant or needs to be inseminated again. When she is pregnant, she is dried off 6 to 8 weeks before her calving date. Ideally, 12 to 14 months after calving, the cow calves again and starts her next lactation.

So, a good detection of oestrus is the first step in getting a cow pregnant in time. Oestrus is the period that a cow is fertile and receptive to the male. Visual detection of oestrus is a challenging job. The oestrus period, which lasts on average 12 hours,

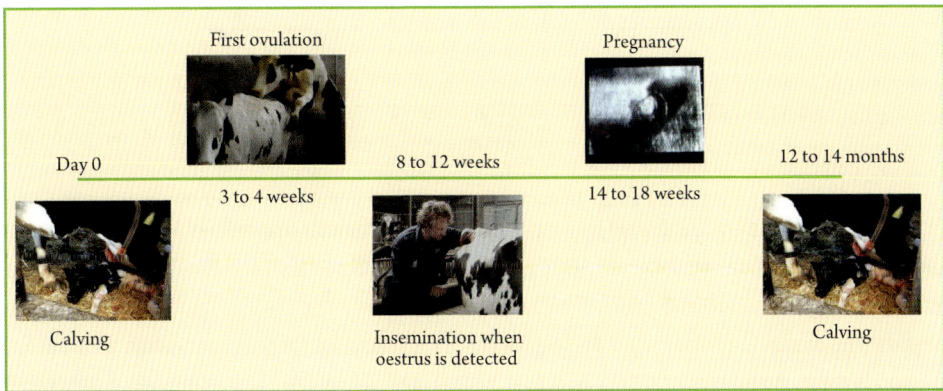

Figure 2.8. Timeline from calving to calving (photos: HAS University of Applied Sciences).

is characterised by distinct behaviour, such as standing to be mounted, restlessness, mounting other cows and a decrease in eating and rumination time. The sensitivity of oestrus detection by visual observation is often only 50% or less (Roelofs *et al.*, 2010). The behavioural changes during oestrus can be detected by sensors.

Many devices for the detection of oestrus are commercially available. Most of them are tri-axial accelerometers (Knight, 2020). All accelerometers are 'at cow', however, the place where the sensor is located differs between sensors. Sensors are mounted on the neck, the leg or in the ear. Also, sensors in boluses that are placed in the reticulum are available for oestrus detection (Table 2.2). It is not the activity *per se* that identifies oestrus by accelerometers, it is the change in the activity of the individual cow from day to day.

The principal of oestrus detection by accelerometers is detecting the increased activity a cow shows during oestrus. The accelerometer registers the activity of an individual cow during a certain period and compares that activity with the expected activity of that specific cow. When the activity exceeds a certain threshold an oestrus alert is generated (Figure 2.9). The performance is determined by how well the device can distinguish between a cow's normal activity and increased activity caused by oestrus. An investment in accelerometers for oestrus detection is likely to be

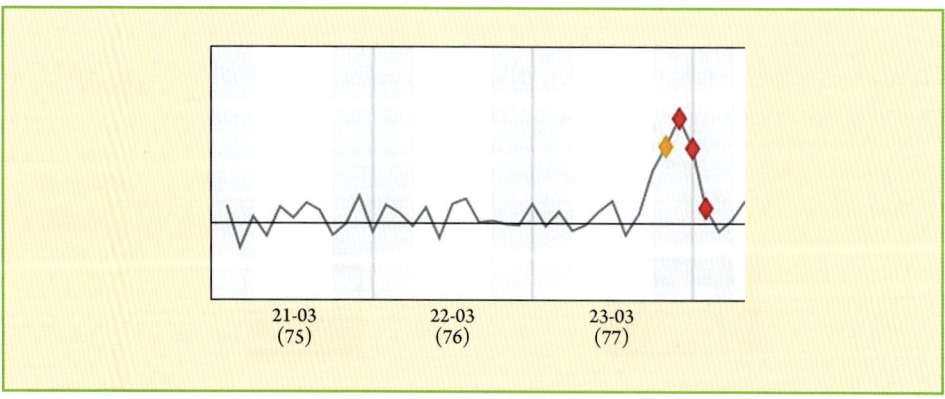

Figure 2.9. Oestrus alerts generated by a leg-mounted accelerometer (example of Nedap CowControl system of a practical training farm of HAS University of Applied Sciences). The dark line represents the relative activity of an individual cow. An algorithm calculates whether the change in activity indicates oestrus. The orange diamond means the relative activity is suspicious for being an oestrus alert, the red diamonds mean the relative activity is an actual oestrus alert. A light grey area is the period between 6 PM and 6 AM and a vertical line is midnight.

profitable for most dairy farms. This strongly depends on the increase in sensitivity that accelerometers achieve, as compared with visual oestrus detection. For different scenarios, the return on invested capital was 10%. The investment was paid back in 8 years for an average farm (Rutten *et al.*, 2014).

The sensitivity and positive predictive value (PPV) of oestrus detection devices differ. The sensitivity, i.e. how many of all the cows in oestrus are detected, is always higher when using an accelerometer compared to visual observation performed a couple of times per day by the farmer (reviewed by Roelofs and Van Erp-van der Kooij, 2015 and Saint-Dizier and Chastant-Maillard, 2012)). The positive predictive value (PPV), i.e. how many of the oestrus alerts correspond with cows being in oestrus, is also important for the performance of an oestrus detection device. The PPV often is less for oestrus detection devices compared to visual observation (reviewed by Roelofs and Van Erp-van der Kooij, 2015). The threshold at which an oestrus alert is generated by an oestrus detection device has a great impact on the sensitivity and the PPV. Physiological factors also play a role in the performance of a device. Factors that influence the expression of oestrus will also influence the sensitivity of an oestrus detection device. Examples of cow-factors are lactation number, body condition score, milk production and lameness. Examples of environmental factors are slippery floors or overstocking (Roelofs and Van Erp-van der Kooij, 2015). Integration of features other than an increased activity such as location and machine learning technologies could optimise sensitivity and PPV (Wang *et al.*, 2020).

Next to accelerometers 'at cow', there are also sensors for oestrus detection 'from cow', e.g. a biosensor that detects the hormone progesterone in milk samples. A biosensor system consists of a sensor, a system to interrogate it and a microcomputer to convert the signal into a format to be displayed to an operator or computer program (Mottram, 2016). The shape and features of progesterone profiles are linked with fertility characteristics and are often considered as 'golden standard' for oestrus. Therefore, insight into the progesterone profile of individual cows could be used to aid the farmer in getting a cow pregnant in time. One system that measures the level of progesterone in the milk regularly during milking is already on the market since 2008 (Herd Navigator, Delaval, Table 2.2). Scientific studies report sensitivity of oestrus detection of this system to be 94 to 99% (Mottram, 2016). This system not only aids in oestrus detection but also gives insight into pregnancy, early embryonic loss and abnormal ovarian structures. Equations are being developed that use the progesterone profile features along with cow-specific traits that can predict the likelihood of insemination success at the time when oestrus is detected. Such an equation can be implemented on farms that monitor progesterone and can support the farmer in deciding whether to inseminate or not (Blavy *et al.*, 2018).

The second step in getting a cow pregnant is insemination at the correct time relative to ovulation. The optimal time for insemination is 12 to 24 hours before ovulation (Roelofs *et al.*, 2006a). To be able to give accurate insemination advice based on oestrus detection devices, the parameters that are measured to indicate the onset of oestrus should have a strong correlation with the time of ovulation and should be consistent between animals. Research has shown that an increase in activity measured by accelerometers can be used to predict the time of ovulation fairly accurate and therefore aid the farmer in deciding when to inseminate a cow (Roelofs *et al.*, 2005). An advantage of a sensor over visual observation is that the start of oestrus is known. With visual observations, it is always a guess when oestrus commenced. Therefore, the optimal time of insemination can be determined more accurately using accelerometers. Most oestrus detection systems that use accelerometers generate not only an oestrus alert but also advice for the optimal insemination moment (Figure 2.10).

Monitoring of progesterone concentrations in the milk alone is less accurate in predicting ovulation time because of the high variation in progesterone profiles between animals (Roelofs *et al.*, 2006b).

Calving

As well as getting a cow pregnant, the prediction of calving time is also a key element in dairy farming. In many systems in north-western European countries, the pregnant

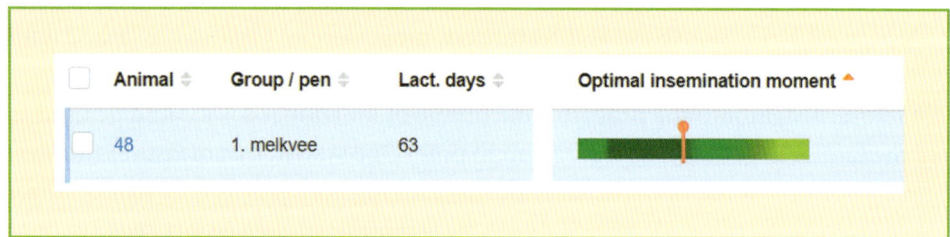

Figure 2.10. Example of information on a dashboard for an oestrus alert and the optimal insemination moment. In this example, cow 48 has an oestrus alert. The green bar represents the optimal insemination moment. When the orange rod is in the dark green area the farmer knows it is the best time to insemination (examples of Nedap CowControl system of a practical training farm of HAS University of Applied Sciences).

cow is dried off six to eight weeks before her due date. During this period a cow does not produce milk and the mammary tissue can recover and repair before the next lactation starts. A dry cow group is a separate group of animals and in many systems, a cow is moved individually to a maternity pen right before calving. Prediction of calving can be used to move her in time and can also be used to decide when human supervision or assistance is required in the calving process (Saint-Dizier and Chastant-Maillard, 2015).

Dystocia is delayed or difficult parturition. Up to one-third of calves born on dairy farms are born after dystocia and dystocia is the cause of over 50% of calf deaths occurring around calving (Rutten et al., 2017). Calving time prediction is thus of critical importance for livestock profitability and animal welfare (Saint-Dizier and Chastant-Maillard, 2015). Calving is characterised by hormonal, behavioural and physical changes. The process of calving is divided into two stages and each has practical and managerial significance. The first stage is represented by dilatation of the cervix and changes in behaviour, such as a reduction in time spent ruminating, more changes between standing and lying and overall more activity. The second stage is the expulsion of the calf. The water bag is ruptured and powerful contractions serve to expel the calf (Scott, 2010). Sensors can alert the farmer for the first and/or second stage of calving (Saint-Dizier and Chastant-Maillard, 2015).

A study in which activity, rumination and ear temperature was used to predict calving time showed that sensor data can be of added value for a more accurate prediction of the start of calving than the expected calving date alone (Rutten et al., 2017). However, the number of false-positive alerts was relatively high and at best, the moment at which second stage calving started was correctly predicted for fewer than half of the calving events. It was concluded that this kind of sensor data might still have merit for the prediction of calving as most alerts were generated within 12 hours before actual calving (stage two) started. Multiple alerts within a few hours for a cow can indicate that the cow should be supervised more closely (Rutten et al., 2017).

Several sensors are available for calving alerts (Table 2.2). These sensors are all 'at cow' and alert the farmer for stage one and/or stage two. The sensors inserted in the vagina measure changes in temperature which indicate stage one of calving with sensitivities of 60 to 70% and positive predictive values of 50 to 70% (reviewed by Saint-Dizier and Chastant-Maillard, 2015). Some of these sensors placed in the vagina also measure the expulsion of the sensor from the vagina, indicating the start of stage two of calving. The sensors mounted on the tail base are inclinometers or triaxial accelerometers that measure the tail raising of a cow which indicates stage one of calving has started. Studies with tail-mounted sensors found sensitivity from 19 to 88% and specificity of 40 to 96% depending on the interval preceding calf expulsion (reviewed by Saint-Dizier and Chastant-Maillard, 2015). It seems a promising sensor for alerting the farmer of calving events. A point of attention is the mounting of the sensor. A study that used a commercially available tail-mounted sensor found that only in 14% of the animals the sensor remained on the tail for the duration of the entire study. The sensor generates an alert when it has fallen off the tail, so it can be placed on the tail of the cow again by the farmer. In 20% of the animals, the sensor had to be removed or was not mounted back when it fell off because of a swollen or painful tail caused by the sensor (Voß et al., 2021). Another 'at cow' sensor that predicts calving is a reticulorumen bolus. It measures temperature changes and gives the farmers notifications of calving on average 15 hours before calving (SmaXtec, 2021).

It is evident that behaviour changes around the time of calving and (in research settings) changes in rumination time, standing and lying behaviour and the number of steps can predict calf birth (Fadul et al., 2017; Titler et al., 2015). However, commercially available accelerometers mounted on the leg, neck or ear do not yet generate calving alerts. An explanation for this is that the management and logistics of each farm can differ substantially when it comes to calving management (grouping and regrouping, individual calving or calving in a group, etc.). This makes it difficult to generate an overall algorithm to correctly identify changes for individual animals to alert the farmer for imminent calving.

2.2.3 Precision health management

The identification of sick dairy cows in an early stage of disease by observations of the individual cow's behaviour has become more and more important with increasing herd size. The standard approach for this task is direct visual observation, which traditionally serves as a diagnostic tool for farmers and veterinarians (Dittrich et al., 2019). In recent years, the use of sensors is becoming increasingly important for animal health management. With an increasing number of cows per herd, it is important that a farmer can identify timely and correctly the cows that need attention. Instead of relying solely on farmers' senses and knowledge, sensors 'at cow', 'from cow' and 'near cow' can provide reliable information about the health or health risks in a herd or of an individual cow. In dairy farming, most technology is used to provide information about individual cows in contrast to poultry and pig farming where technology is mostly used to give information about a flock or group of animals rather than the individual animal (see Chapter 4 and 5). Many devices are available that generate health alerts (Table 2.2).

'At cow' sensors that provide information about the health of a cow are mostly accelerometers, measuring behaviour such as rumination, eating and/or lying. The farmer receives an alert when changes on individual cow level occur. Activity and feeding behaviour patterns play a major role in cow health. These patterns consist of various characteristics, which interact closely with each other. That is why they are often part of behaviour-based health monitoring (Dittrich *et al.*, 2019). Which behaviour or characteristics are being measured to which a health alert is generated depends on where the device is mounted. Sensors mounted on the neck measure rumination and/or eating behaviour. Sensors mounted on the leg measure standing, walking and lying behaviour, sensors mounted on the ear measure all the mentioned behaviours and sensors placed in the reticulorumen measure general activity without distinguishing which behaviour. Some devices also have a temperature sensor beside the accelerometer (ear tags or reticulorumen boluses). This is useful as changes in temperature can be a result of health problems (Lukas *et al.*, 2008). A new characteristic that is studied but not yet implemented is sensor information about drinking behaviour. Monitoring drinking events with a temperature sensor in a

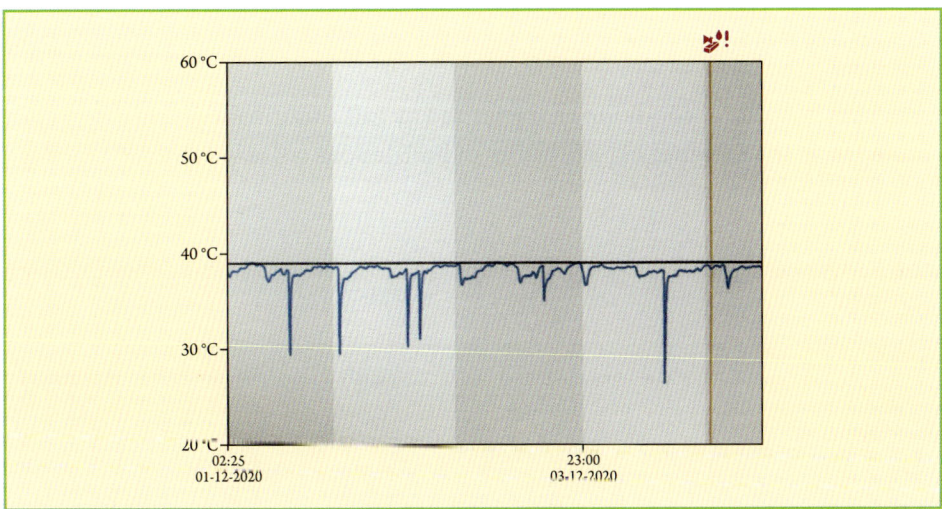

Figure 2.11. Example of visualisation of drinking behaviour of a cow. A spike represents a drinking event. This cow had an alert of reduced drinking behaviour (red icon, example of an alert with the SmaXtec system on a practical training farm of HAS University of Applied Sciences).

reticulorumen bolus could be used for health prediction. When a cow is drinking, the reticulorumen temperature shows a sudden drop (Figure 2.11). A sudden increase in drinking events may be associated with a substantial change in ambient temperature and may indicate heat stress (West, 2003). It was also shown that e.g. cows suffering from acute mastitis spent less time drinking (Vázquez-Diosdado et al., 2019).

It is clear that accelerometers are very helpful in detecting changes in behaviour. These sensors are accurate, can detect deviations earlier than humans and measure 24 hours and seven days per week. However, a skilled person (farmer, veterinarian or adviser) is essential in making a diagnosis of what is the cause of the change and deciding which action to take. Different health issues can result in similar changes in behaviour. In other words, a particular change in behaviour can be caused by different health issues. This is illustrated in Table 2.3.

So when a health alert is given based on a change in activity, the farmer knows which cows to check. Most sensor systems show on a dashboard which cows need to be monitored, based on abnormal behaviour. This makes accelerometers a useful tool to aid the farmer in deciding which cows need attention at which times. In practice, a farmer can check the dashboard several times a day and inspect the animals that are on the list for example during milking. To be even more useful, sensors should not only indicate that something may be wrong but also what is predicted to be wrong. In the next sections, the use of sensors to detect different health issue are discussed.

Lameness

Lameness is a painful condition characterised by gait abnormalities and discomfort coming from the presence of painful foot or limb lesions (Alsaaod et al., 2019). It is a

Table 2.3. Changes in behavioural patterns for different health issues (adapted from a review by Dittrich et al., 2019).[1]

	Hypocalcaemia	Ketosis	Metritis	Mastitis	Lameness
Standing	↑	↑↓		↑	
Lying	↑		↑	↑↓	↑
Feeding	↓	↓	↓	↓	↓
Ruminating		↓	↓	↓	↓
Physical activity		↓	↓	↑↓	↓

[1] ↑ = an increase in duration is found in research; ↓ = a decrease in duration is found in research.

severe welfare problem and a production-limiting disease in dairy farming. Lameness management consists of both prevention and treatment. For a lame cow to be treated, it must first be identified as lame by the farmer, this generally occurs in three ways. The first is using a locomotion scoring system to assess a herd systematically. The second is routine hoof trimming. Here, legs are lifted, inspected, and if required, treated. The third, and most common, is ad hoc observation during other activities, such as herding or cleaning stalls (O'Leary et al., 2020). Regular locomotion scoring or hoof trimming is time-consuming and depends on skilled persons. It was shown that dairy farmers are aware of only one-quarter of lame cows in their herd (Fabian et al., 2014). Early lameness detection and prompt treatment reduce the duration and prevalence of lameness and thus improve cow welfare and farm profitability. Detection of moderately lame cows (underestimated even by trained observers) is the key for early intervention to prevent deterioration of claw disorders (Alsaaod et al., 2019). Improvements in the diagnosis of lameness are likely to result in earlier treatment and improved clinical outcomes.

Automatic lameness detection systems are developed to replace locomotion scoring by measuring various traits of locomotion and/or behaviour for application in practice or research purposes, using different types of sensors techniques (Alsaaod et al., 2019). Several 'near cow' sensors are being developed or on the market to study the movement of the cow's body, e.g. video systems and walk-over sensors. Video systems make use of analysis of cows walking to measure changes in several features that indicate lameness. Features that are analysed are gait characteristics, stride length, stride duration, average speed and step height, back posture and curvature and movements of the spine and hind limbs. Scientific studies have shown an accuracy for detecting lame cows of around 95% using video analysis (reviewed by Alsaaod et al., 2019). However, these sensor systems are still in the research stage and no applications are commercially available yet.

Walk-over systems are also 'near cow' sensor systems that can be used to detect lameness. One of these systems is a system that measures ground reaction forces exerted by the cows while walking (StepMetrix™). Certain limb movement variables are derived from these ground reaction forces. Changes in limb movement can be caused by lameness. The use of different models led to sensitivities between 70 and 84%, specificities between 69 and 80% and positive predictive values between 49 and 65% (Wu et al., 2011). This system is commercially available.

Another walk-over system is a pressure-sensitive walkway (GaitWise system). The sensor is a pressure-sensitive mat placed on the floor in a corridor where the cows pass once or twice a day. Daily measurements are available of the placing, timing and relative pressure of the cows' hooves on the ground. From this, several variables can be calculated for use in a detection algorithm to alert the farmer if a cow is lame. The pressure sensors detect stride-to-stride fluctuation which are gait inconsistencies associated with lameness (Figure 2.12).

The model used to detect mildly lame cows had a sensitivity of 88% and a specificity of 87% (Van Nuffel et al., 2015). Since the GaitWise pressure mat is based on an existing sensor originally intended for application in human medicine, the sensor provides very detailed information at a high cost. This cost is too high to justify the investment for most dairy farms. Because cows need to walk normally during monitoring, enough space (2 m) before and after the sensor is required, resulting in a total minimum length of about 10 m. Therefore, making the system smaller and more compact would also be beneficial for its adoption in practice. A study has found that downscaling the length of the mat and sensor resolution was possible without

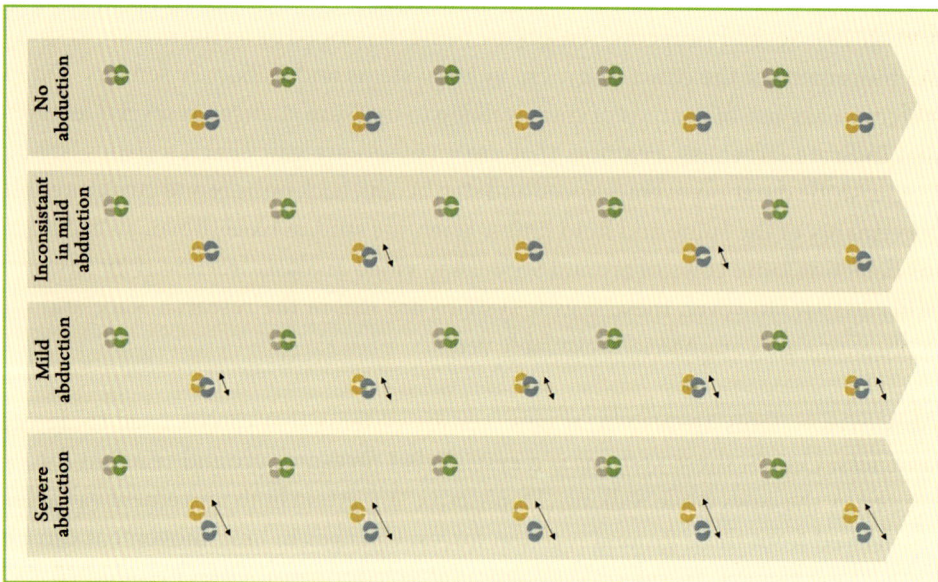

Figure 2.12. Visualisation of the gait inconsistency hypothesis for early lameness detection: before non-lame cows show mildly or severe abduction during the development of lameness, they first show stride-to-stride fluctuations or 'inconsistency' in showing the mild abduction (Van Nuffel et al., 2015).

significant loss in lameness detection. Future research is recommended to quantify the changes in cost when comparing the original sensors system to this downscaled alternative (Van De Gucht et al., 2017). Up to now, the system is not commercially available.

A walk-over system using ultrasonography is another way of detecting lameness. In this system, cows are guided to stand on a hoof scan system. The hoof scan takes an ultrasonography of the hooves from underneath while the animals are standing in a pool of freshwater to prevent air from being trapped between the sole-horn and the floor. The image displays the space between the floor and sole horn and thus whether or not it corresponds with the ideal model. Next, the bodyweight of the cow is recorded using the integrated weighing system. The images and weight measurements are automatically analysed by the system, determining possible deviations from the physiological shape. The data is automatically transmitted to the local management system of the farm. At the next scheduled appointment for hoof maintenance, cows with deviations will be set apart in the hoof care box area (Fiedler et al., 2019). This system is not yet on the market and performance in terms of sensitivity or PPV of the system in detecting lame cows is not yet assessed.

The use of infrared thermography (IRT) to identify lameness steadily increased in recent years largely because of its non-invasive properties, ease of automation and continuing cost reduction. Infrared thermography is a non-invasive technique that measures emitted infrared radiation and displays the information pictorially as a thermogram (colour-scale) of the surface temperature of an object. A primary reason for using IRT to identify lameness is that if lameness is due to a localised inflammatory process, a thermal signal is expected to occur at the respective site. Although thermography may not identify the specific type of lesion, it assists in defining the area of increased inflammation and/or injury (Alsaaod et al., 2019).

'At cow' sensors such as accelerometers are used to investigate whether differences in gait characteristics that can be measured by acceleration are useful to detect lame cows. For example (a)symmetry of variance of the hind leg during forward acceleration and differences in limb acceleration are studied. They show good potential for correctly identifying lame cows in an early stage (Alsaaod et al., 2019). A study that used leg mounted accelerometers to investigate whether this technology may help farmers to detect lame cows concluded that the activity level, combined from different variables calculated from walking and standing, decreased already from healthy cows to mildly lame cows (Thorup et al., 2015).

Although many prototypes of automatic lameness detection systems for dairy cows exist, the majority of these systems are still in the research or development phase and have not yet been commercialised. One system with integrated decision support that has been implemented under field conditions is CowAlert from Icerobotics. It is the first accelerometer-based automated lameness detection system that has been marketed in 2017 (O'Leary *et al.*, 2020). Daily Lameness Alerts from this system provide a quick overview of the lameness status of the herd based on colour codes and individual cow lameness history allows a farmer to examine the lameness status of each cow (Figure 2.13). This information helps to reduce the number, severity and duration of lameness cases (IceRobotics, 2021).

Mastitis

Mastitis is an inflammation of the udder or mammary gland. It is commonly classified into subclinical, clinical, and chronic forms, all of which cause significant animal welfare concerns. Subclinical mastitis is the presence of infection without apparent signs of local inflammation or systemic involvement. Although transient episodes of abnormal milk may appear, subclinical mastitis is, for the most part, asymptomatic. If the infection persists for at least two months, the infection is termed chronic. Once established, many of these infections persist for entire lactations or the life of the cow. Clinical mastitis is an inflammatory response to infection which causes visibly

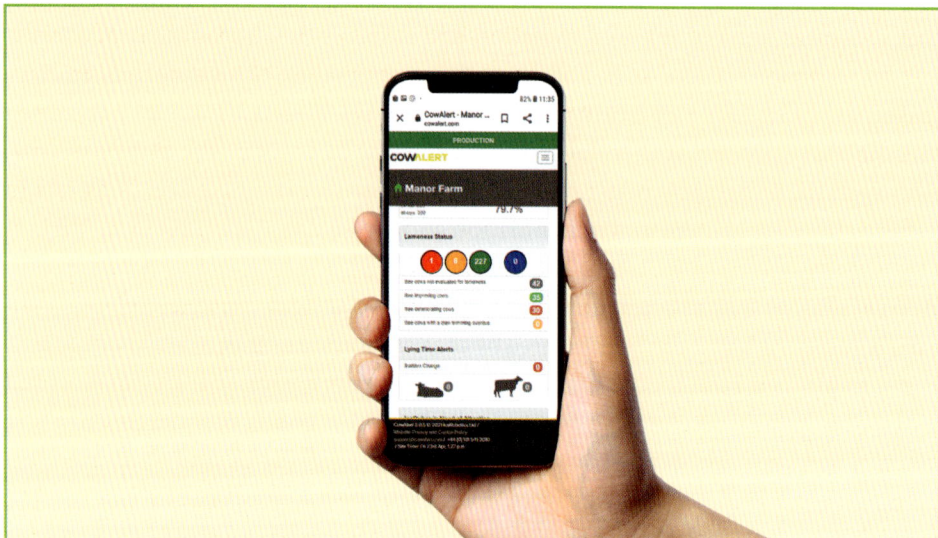

Figure 2.13. Example of visualisation of lameness attention on a smart phone (courtesy of IceRobotics, Scotland UK).

abnormal milk (abnormal colour, fibrin clots). As the extent of the inflammation increases, changes in the udder (swelling, heat, pain, redness) may also be apparent. Clinical cases that include only local signs are referred to as mild or moderate. If the inflammatory response includes systemic involvement (fever, anorexia, shock), the case is termed severe. If the onset is very rapid, as often occurs with severe clinical cases, it is termed acute or severe mastitis. Although any number of quarters can be infected simultaneously in subclinical mastitis, typically only one quarter will display clinical mastitis (Erksine, 2020).

The sensors used for mastitis detection are mostly 'from cow' sensors which measure features of the milk of individual cows. The main sensor system for the automated detection of mastitis is electrical conductivity (EC) of the milk (Rutten *et al.*, 2013). The EC is determined by the concentration of anions and cations. If the cow suffers from mastitis, the concentration of Na^+ and Cl^- in the milk increases, which leads to increased EC of milk from the infected quarter (Norberg *et al.*, 2004). EC sensors are sometimes combined with milk colour sensors. These sensors are in-line sensors, meaning that they measure in a continuous milk flow. Most automatic milking systems have EC and milk colour sensors incorporated for measuring the features during milking. Detection performance varies considerably between studies. Studies have found sensitivity between 55 and 89% and specificity between 56 and 99%. A trade-off exists between sensitivity and specificity as high sensitivity (>80%) is combined with low specificity (reviewed by Rutten *et al.*, 2013). Even if systems do not have high performance in terms of sensitivity concurrent with specificity, the value that an automated mastitis detection system provides to a farmer is obvious, as the alternative is having no automated detection at all.

Another sensor system used for mastitis detection is a biosensor that detects certain enzymes associated with mastitis (Herd Navigator, DeLaval). A commercially available biosensor automatically takes representative milk samples from specific cows of the herd during milking and analyses parameters on-line that help farmers to monitor mastitis, among other things. The system contains unique biological models, which take into consideration the measured parameters, cow information and additional risk factors to keep the herd healthy. For the detection of mastitis, lactate dehydrogenase (LDH) is measured using a colorimetric principle. The system can detect clinical and subclinical mastitis up to 3 to 4 days before clinical signs are visible in the animals affected. The sensitivity of the system is more than 80% (Mazeris, 2010).

Detection of subclinical mastitis without sensor systems is predominantly done by testing milk for somatic cell counts (SCCs) (predominantly leukocytes) using either the California Mastitis Test on the farm or automated methods provided by dairy herd improvement organizations on, for example, monthly intervals. SCCs are positively correlated with the presence of infection. Inflammatory changes and decreases in milk quality may start with SCCs as low as 100,000 cells/ml. Although variable (especially if determined on a single analysis), SCC of ≥200,000 cells/ml in a cow indicates a high likelihood of infection (Erskine, 2020). SCC sensors are available that measure on-line during each milking the somatic cell count (Table 2.2). Sensitivity and specificity can reach up to 80 and 90%, respectively (Dalen *et al.*, 2019).

A 'near cow' sensor that is on the market for the detection of mastitis is a thermal imaging or infrared camera (AgriCam CaDDi Mastitis). Infrared thermography is an effective method that shows physiological changes, so it is useful for diagnosis and pain assessments. Inflammatory processes without pronounced swelling can result in detectable increases in surface temperature. Inflamed tissues show changed circulation and one of the signs of inflammation is the heat that arises from local vasodilatation (McManus *et al.*, 2016).

The infrared camera sensor system is a thermal imaging system that provides a visual representation or 'heat-map' of the cow's body and shows a real-time distribution of the heat (Figure 2.14).

The cameras are placed at the entrance of the milking parlour. Images collected by the cameras are automatically analysed by software that uses algorithms to identify temperature fluctuations. If anomalies are detected, the software will notify the farmer. The sensor system can detect mastitis two to three days before the onset of clinical signs. Sensitivity or other performance values are not known (French, 2016).

2.2.4 Precision feeding management

Feeding behaviours are important for the dairy cow, so nutritional management is one of the main and most important activities in dairy farming, as it impacts milk production and animal welfare. Also, feed costs represent 40 to 60% or even more of the total cost of production, thus reinforcing the need for adequate planning and control of feeding. Optimising feeding management requires a careful analysis of various animal-related data, such as age, weight, milk production, and stage of lactation (Da Rosa Righi *et al.*, 2020). The ration of a cow in an intensive dairy

2. Precision dairy farming

Figure 2.14. A thermal image from an infrared camera for the detection of mastitis (photo: courtesy of Agricam, Sweden).

farming system consists of roughage and concentrates, often supplemented with by-products. In most systems for feeding dairy cows, roughage is fed *ad libitum* and the ration is enriched by adding concentrates to the roughage mixture on herd or group level and/or by supplying extra concentrates to the individual dairy cow. Standard guidelines for feeding are such that the amount of energy offered is in line with the energy requirement for maintenance, pregnancy and production. Within this system, there is uncertainty both on the requirement and on the intake side (André, 2011).

Computer-controlled self-feeders are developed to allocate concentrates to individual cows according to their needs. The first individual feeding technology was developed to feed concentrates individually according to the farmers' decision. The technology was designed to correlate the amount of concentrates fed to the cow according to her production. The energy requirement for production is derived from recorded milk yields, i.e. past performance, but there is uncertainty about the expected milk production in the future. Furthermore, there is uncertainty in the prediction of the requirement for maintenance, because body weight and weight change are usually unknown in practice (André, 2011).

To be able to feed cows concentrates according to their individual needs, sensors are required. A combination of sensors is used to get optimal concentrate rationing. Ideally, these sensors should measure milk yield, milk composition, body weight, body composition and feed intake (Maltz, 2020). However, the on-farm measurement of individual feed intake, especially forage or total mixed ration is not feasible. This system combining different sensor data is not yet commercialised. The dynamic feed system that is used in practice derives daily individual settings from the actual individual milk yield response to concentrate intake. This response is estimated using an adaptive dynamic linear model. Optimal daily individual settings for concentrate supply are directed to achieve the maximum gross margin milk returns minus concentrate costs. This response curve plays a key role in the application of dynamic feeding, and the idea is that every cow reacts differently on extra or less concentrates (André, 2011). It was stated that this system may help to find the most efficient way to feed an individual cow. The system is very applicable to limit (too) high concentrate levels for late lactation cows because most farmers still feed their late lactation cows concentrates when they do not need the concentrates anymore. In the dynamic feeding model, there is no monitoring of (rumen) health parameters which may be a risk factor for excessive concentrate intake and related diseases (Meijer and Peeters, 2010). In the future, this could be a valuable addition to dynamic feeding.

Sensor systems can also be used to optimise feed delivery at the feed bunk when using automatic feeding systems (AFS). Several different AFS are on the market. All AFS can mix and distribute the feed automatically and some AFS can also fill the mixer automatically (Figure 2.15, DLG Committee for Technology in Animal Production, 2014).

AFS gain in importance due to rising demands on performance-related feeding of cows and animal welfare. But so far, under practical conditions, AFS are used with static settings for feeding times and frequencies without taking into account animal demands. A study investigated whether it was possible to develop a method for feeding cows with an AFS at dynamic times concerning their behaviour (Oberschätzl-Kopp et al., 2016). Real-time location data of the cows was used to register feeding behaviour. It was concluded that dynamisation of feed delivery times according to cows duration of stay at the feed bunk could be beneficial for even distribution of feed bunk visits and meals throughout the day. AMS milkings were not evenly spread throughout the day when feeding times were controlled dynamically. The present implementation of real-time processed positioning data is a possible approach for combining animal- and feed-related data of different sensor technologies to advance AFS for feeding cows according to their needs. Based on this information, recommendations for farmers regarding the AFS settings of feeding times and amounts of feed can be derived. This type of dynamic automatic feeding in which the times of feed delivery

Figure 2.15. An automatic feeding system that fills the mixer (left picture) and mixes and distributes the feed automatically (right picture) (photos: courtesy of Lely).

is adapted to the cows' behaviours instead of fixed times for feed delivery is still in the research phase.

2.2.5 Precision youngstock management

Livestock management based on sensor systems has been adopted on many dairy farms around the world, but these systems have traditionally been used with adult animals, not calves. As already described, the first sensor application was for the detection of oestrus in dairy cows, which is a feature that does not apply to young calves. With the development of the use of sensor systems to aid in health and welfare management, it is becoming an interesting area to see which systems can be used in youngstock rearing as well. Sensors systems can be used for managing the performance and health of (pre-weaned) calves (Costa *et al.*, 2021).

Tri-axial accelerometers mounted on the leg, neck or in the ear are the predominant 'at cow' sensors used in calves. Research has shown that these sensors can be used to measure lying behaviour, step activity and locomotor play accurately (over 90% sensitivity and specificity). Validation of accelerometers to measure rumination and feeding behaviour has had mixed results. The differences between studies suggest that further technology development is needed to reliably measure rumination in calves (reviewed by Costa *et al.*, 2021).

The accelerometers that are used to detect health issues in adult cows, may also be useful in detecting health problems in calves. Calves suffering from bovine respiratory disease (BRD) were less active before, on the day of, and after respiratory disease diagnosis. Furthermore, diseased calves had reduced lying frequencies starting two days before diagnosis, as well as after diagnosis (Swartz *et al.*, 2017). Step activity and lying bouts may prove to be a useful tool in identifying respiratory disease under practical farming, but this requires further research (Duthie *et al.*, 2021; Swartz *et al.*, 2017). As described for dairy cows, different health issues can result in similar changes in behaviour in calves. In other words, a behaviour change can be caused by different health issues. For example, navel inflammation was associated with less lying time and diarrhoea was associated with reduced lying times two days before and increased lying time on the day of diagnosis (Costa *et al.*, 2021). Disease likely affects activity in calves that can be measured by accelerometers. At the moment, it is up to the farmer (or veterinarian) to diagnose the problem when a change in behaviour is detected by the sensor system. In the future, it might be possible to diagnose the health issue automatically by the sensor system. This requires further research.

A 'near calf' sensor system is the automated feeding system (AFS). Automated calf feeders have gained popularity as a calf-management tool and are extensively used (Costa *et al.*, 2021). They have the versatility of administering nutritional plans (i.e. milk and concentrates) and can be used to monitor the feeding behaviour of individual calves. Some AFS can monitor and record individual feeding behaviour and/or weight of dairy calves. The AFS technologies record daily milk or solid feed intake, drinking speed per minute (milk), and visits to the feeder (rewarded with milk or solid feed, or unrewarded without milk or solid feed), and can be very useful for the management of dairy calves. It was found that changes in feeding behaviour, such as drinking speed, the number of unrewarded visits and intake can be an early warning system for BRD or diarrhoea (Costa *et al.*, 2021). A combination of feeding behaviour measured by an AFS and lying behaviour measured by accelerometers was found to be useful in developing algorithms that provide farmers with an automated alert system to identify calves at risk (Lowe *et al.*, 2021).

Other technologies that are under investigation for calf management are sensors that can measure body temperature ('at calf' in the ear, implanted as a microchip in the neck, or in a rumen bolus or 'near calf' with infrared thermography), heartrate, ('at calf' with monitors strapped around the thorax) and weight and body condition ('near calf' with cameras or weighing scales). Further software development is necessary to scale these technologies to automatic recording on commercial farms (Costa *et al.*, 2021).

2.2.6 Precision grass management

Precision technologies in grass-based dairy systems

The desire to know and maybe even control the variability in grass allowance forms a good starting point for introducing precision technologies. However, the introduction of precision farming on grass-based dairy farm systems has been limited in the recent past. There are several reasons for this (Shalloo *et al.*, 2018). First, the nature of grass-based systems with large grazing platforms comes with corresponding connectivity challenges and associated costs. Second, grass-based farm systems are often low-input and low-cost systems, which lead to limited investments in general and thus also in precision farming. The precision technology that has been described in the previous section of this chapter (PLF in the farm; precision dairy management) is relevant for grass-based systems as well. This section adds precision technology that is meant to specifically support grass-based systems. The section is divided into technology related to (1) animals and (2) grasslands.

Precision technology related to animals in grass-based dairy systems

Accelerometers, which are often used to detect cow behaviour, such as movement, rumination and resting, can also fulfil useful functions in grassland management. These sensors are usually positioned on the head, neck, ear or leg of a cow. Due to this specific position, they can measure the grazing activity of a cow. Grass intake is a key factor in grass-based dairy systems. The profitability of the farm highly depends on sufficient grass intake. As mentioned before, the grass allowance is highly variable. Monitoring grazing behaviour of individual cows can therefore support grazing management. Grazing cows have different feeding behaviour than cows kept inside. Therefore, available sensors to measure feeding behaviour indoors are not automatically relevant outdoors and may need to be adapted. In the last few years several sensors, usually head/neck sensors, have been developed to effectively monitor grazing behaviour. Some examples are:

- RumiWatch (based on jaw movements; it can detect rumination chews and grazing bites and is primarily used for research) (Werner *et al.*, 2018).
- Heattime (Molfino *et al.*, 2017).
- Nedap (Dela Rue *et al.*, 2020).

Accelerometers can also detect lameness (see Section 2.2.3). Preventing lameness is especially important in grass-based systems since cows have to walk considerable distances from the field to the stable to be milked and vice versa.

Global position system (GPS) based tracking systems can be used to localise cows (Williams *et al.*, 2016). The application of GPS in virtual fencing systems might be a potentially interesting feature. Virtual fencing aims to remotely map and control grazing paddocks without the use of fixed fences. This is achieved by cow sensors that register the position of the cow in the field. Virtual fences are defined by GPS coordinates and cows receive signals once they come too close to these virtual fences. Grass-based dairy systems are depending on frequent and precise allocations of grass to animals. Allocating the correct amount of grass requires a considerable amount of labour. A virtual fence system would allocate the required area to animals without the associated labour. Such a system depends on GPS to determine boundaries. At the moment, the system only works reliably with associated visual cues that guide the cows throughout the system (McSweeney *et al.*, 2020). A simple alternative to reduce labour when moving grazing cows from paddock to paddock is to work with timed and remote release gates that enable grazing cows to move through the gate from one paddock to another at a predestined time.

Precision technology related to grasslands

Insight into the available grass herbage mass is essential to optimise the grazing system. Measuring of grass can be done using a simple rising plate meter, but currently, there are also options to automate aspects of collecting the data, e.g. via digitally enabled mechanical plate meters or ultrasonic sensor-based plate meters (Shalloo *et al.*, 2018). The latter plate meters use ultra/micro-sonic sensors that measure the distance to the height of the plate resting on the canopy (Moeckel *et al.*, 2017). Some meters, like the C-Dax Pasture Meter, use infrared light for this purpose. New and innovative rising plate metres, like the so-called Grasshopper (McSweeney *et al.*, 2019), use GPS to also automatically record the position of the readings.

Spectral analysis (which can either be terrestrial or remote) provides opportunities to measure the available herbage mass, for example by using drones or satellites or handheld equipment (Shalloo *et al.*, 2018). Usually, the NDVI (Normalised Difference Vegetation Index) is used to provide an estimate of the available herbage mass, but the results are variable. In general, differences in biomass between parcels can be differentiated, but it is more difficult to provide a reliable estimate of the absolute biomass. The method is challenging, mainly due to atmospheric effects and variations at the field level (variations in height of the pasture, variations in pasture growth, variation in the growth stage, i.e. flowering versus non-flowering plant). Grazing in itself further complicates this, as due to selective grazing, rejected patches occur, that do contribute to standing biomass, but will not be eaten by the cows. Klootwijk *et al.* (2021) presented a method to use multispectral images to correct fresh grass allowance for selective grazing and thus provide insight into the actual availability of fresh grass for the cows rather than insight into the total biomass standing.

Spectral reflectance could also be used to provide indications of spatial variability of minerals like nitrogen in the soil of a particular field. This information could then be used to optimise the fertilisation of grasslands via precision application (Berry *et al.*, 2017).

2.3 Future trends in precision dairy farming

The global value of biosensor technologies for animals is estimated at >16.5 billion euros by 2020 (Neethirajan, 2017). A large share of that will be spent in the dairy sector. In a relatively short time, many sensors for dairy cows have been developed, mainly in the market of oestrus detection. This means that there is potential for much more development in the near future. Health and welfare monitoring might become widespread, in dairy cows as well as in other dairy species. Technologies that can measure parameters such as eating and drinking already exist, and the possibility to extract relevant information from complex datasets is increasing constantly. Dairy farms are expanding in size, which results in a growing need for technologies assisting in animal husbandry. Data also gives insight into animal welfare, and these data can add to the value of the primary product (milk). Sensor systems can benefit the farmer as well as the animals, but also consumers and other players in the dairy foods chain. To be successful, a combined and coordinated effort is needed from biologists, engineers and business experts in areas of R&D and marketing. Sensor-based husbandry support could be an important factor of the dairy sector's vital contribution to global food security (Knight, 2020).

2.3.1 Future trends in modelling and algorithms

The performance of a sensor system depends on its ability to generate correct attentions, correctly steer processes or give the correct insights that lead to the correct actions. The more features are measured by sensors, the more challenging it is to analyse and process this data, to generate useful insights and actions traditionally by programming. Machine learning (ML) will become more and more important to develop these algorithms. ML algorithms are used to automatically learn patterns and make inferences from data (Lokhorst *et al.*, 2019). ML algorithms build a model based on sample data, known as 'training data', to make predictions or decisions without being explicitly programmed to do so (Mitchell, 1997). ML technologies can provide high accuracies for the detection/prediction of e.g. oestrus (Wang *et al.*, 2020), calving time or health issues, such as lameness and mastitis (Slob *et al.*, 2021). A challenge is the standardisation of the collection and sharing of data globally. However, as more farms get connected to technology, artificial intelligence and sensing technologies will start playing a more decisive role in helping farmers see patterns and solutions to pressing problems in future animal farming systems. While there are still several unknowns, limitations, and open-ended questions, one thing is certain: in this decade, we will discover the true power of human-artificial intelligence collaborations in the livestock sector (Neethirajan, 2020).

2.3.2 Future trends in sensor systems

It is widely recognised that the development of precision dairy farming has positive effects on the system and that sustainable intensification of the dairy sector is one of the most important challenges for the near future. Many sensor systems are available that aid the farmer in managing the cows' reproduction and health. Although the beneficial effects of PLF have been studied, the full quantification of the environmental, economic and social sustainability of dairy cattle livestock production equipped with PLF techniques has not yet been carried out. Future studies should be done to quantify these effects and to compare solutions with and without PLF to evaluate the effective sustainability with a life cycle approach adopting Life Cycle Assessment (LCA), Life Cycle Cost (LCC) and Social Life Cycle Assessment (SLCA) methods. This will allow policymakers and stakeholders to make decisions and introduce incentives and policies to promote the introduction of PLF on farms (Lovarelli *et al.*, 2020).

Another very important area in which sensors systems can play a crucial role is the growing demand for higher standards of welfare and reliable methods to measure or assess the mental and physiological wellbeing of animals (Chapa *et al.*, 2020). Animal welfare is a multidimensional phenomenon. To assess animal welfare different assessment have been proposed, such as FAWC 'Five Freedoms', 'Welfare Quality®'

Programme and Mellor's 'Five Domains' Model (Van Erp-van der Kooij and Rutter, 2020). In all these assessments nutrition, environment, health and behaviour play a role.

A range of sensors has been developed to improve the efficiency of animal production by optimising management. Data from these sensors could be integrated into automated welfare assessment systems. E.g. the combination of body condition scoring assessed by a camera, eating and rumination time assessed by accelerometers and drinking behaviour assessed by reticulorumen boluses can be used to assess nutrition. Further research is needed to define and validate this integrated approach (Van Erp-van der Kooij and Rutter, 2020).

2.3.3 Future trends in precision grass-based dairy farming

Currently, public awareness of the importance of ecosystem services is increasing. Next to the production of feed, grasslands provide many other ecosystem services, e.g. biodiversity and carbon sequestration. Van den Pol-van Dasselaar *et al.* (2021) promote new business models where farmers are rewarded for added value. This requires that the associated ecosystem services need to be monitored and measured. Precision technology, for example, remote sensing techniques for monitoring purposes, could support this.

In general, systems with complementary sensors and good decision support applications are needed, which will help to overcome the limitations of single sensors and to provide better information about animals and grasslands.

References

4D4F, 2019. Data driven dairy decision for farmers. Website. Available at: https://www.4d4f.eu/

Agriculture and Horticulture Development Board (AHDB), 2021. UK and EU cow numbers Available at: https://ahdb.org.uk/dairy/uk-and-eu-cow-numbers.

Alsaaod, M., Fadul, M., Steiner, A., 2019. Automatic lameness detection in cattle. Veterinary Journal 246: 35-44. https://doi.org/10.1016/j.tvjl.2019.01.005

André, G., 2011. Adaptive models for operational use in dairy farming. Wageningen University, Wageningen, the Netherlands. Available at: https://edepot.wur.nl/170222.

Augère-Granier, M.-L., 2018. The EU dairy sector: main features, challenges and prospects – think tank. Available at: https://tinyurl.com/v4zxwr4e.

Barkema, H., von Keyserlingk, M., Kastelic, J., Lam, T., Luby, C., Roy, J., LeBlanc, S., Keefe, G., Kelton, D., 2015. Invited review: changes in the dairy industry affecting dairy cattle health and welfare. Journal of Dairy Science 98: 7426-7445. https://doi.org/10.3168/jds.2015-9377

Berry, P.M., Holmes, H.F., Blacker, C., 2017. Development of methods for remotely sensing grass growth to enable precision application of nitrogen fertilizer. Advances in Animal Biosciences 8: 758-763. https://doi.org/10.1017/s2040470017000863

Bewley, J.M., Robertson, L.M., Eckelkamp, E.A., 2017. A 100-year review: lactating dairy cattle housing management. Journal of Dairy Science 100: 10418-10431. https://doi.org/10.3168/jds.2017-13251

Blavy, P., Friggens, N.C., Nielsen, K.R., Christensen, J.M., Derks, M., 2018. Estimating probability of insemination success using milk progesterone measurements. Journal of Dairy Science 101: 1648-1660. https://doi.org/10.3168/jds.2016-12453

Centraal Bureau voor Statistiek (CBS), 2021. Landbouw; gewassen, dieren, grondgebruik en arbeid op nationaal niveau. Available at: https://tinyurl.com/3fcdma3x.

Chapa, J.M., Maschat, K., Iwersen, M., Baumgartner, J., Drillich, M., 2020. Accelerometer systems as tools for health and welfare assessment in cattle and pigs – a review. Behavioural Processes 181: 104262. https://doi.org/10.1016/j.beproc.2020.104262

Costa, J.H.C., Cantor, M.C., Neave, H.W., 2021. Symposium review: precision technologies for dairy calves and management applications. Journal of Dairy Science 104: 1203-1219. https://doi.org/10.3168/jds.2019-17885

CRV, 2021. Jaarstatistieken 2020 voor Nederland. Available at: https://www.cooperatie-crv.nl/downloads/stamboek/publicaties/.

Da Rosa Righi, R., Goldschmidt, G., Kunst, R., Deon, C., André da Costa, C., 2020. Towards combining data prediction and internet of things to manage milk production on dairy cows. Computers and Electronics in Agriculture 169: 105156. https://doi.org/10.1016/j.compag.2019.105156

Dalen, G., Rachah, A., Nã, V., Schukken, Y.H., Reksen, O., 2019. The detection of intramammary infections using online somatic cell counts. Journal of Dairy Science 102: 5419-5429. https://doi.org/10.3168/jds.2018-15295

Delaby, L., Finn, J.A., Grange, G., Horan, B., 2020. Pasture-based dairy systems in temperate lowlands: challenges and opportunities for the future. Frontiers in Sustainable Food Systems 4: 543587.

Dela Rue, B., Lee, J.M., Eastwood, C.R., Macdonald, K.A., Gregorini, P., 2020. Short communication: Evaluation of an eating time sensor for use in pasture-based dairy systems. Journal of Dairy Science 103: 9488-9492. https://doi.org/10.3168/jds.2020-18173

Dittrich, I., Gertz, M., Krieter, J., 2019. Alterations in sick dairy cows' daily behavioural patterns. Heliyon 5: E02902. https://doi.org/10.1016/j.heliyon.2019.e02902

DLG Committee for Technology in Animal Production, Oberschätzel, R., Haidn, B., 2014. Automatic feeding systems for cattle. DLG Expert Knowledge Series 398 DLG Expert Knowledge Series 398. Available at: https://tinyurl.com/788n55df.

Duthie, C.-A., Bowen, J.M., Bell, D.J., Miller, G.A., Mason, C., Haskell, M.J., 2021. Feeding behaviour and activity as early indicators of disease in pre-weaned dairy calves. Animal 15: 100150. https://doi.org/10.1016/j.animal.2020.100150

Erksine, R.J., 2020. Mastitis in cattle. Merck Veterinary Manual. Available at: https://www.merckvetmanual.com/reproductive-system/mastitis-in-large-animals/mastitis-in-cattle.

Fabian, J., Laven, R.A., Whay, H.R., 2014. The prevalence of lameness on New Zealand dairy farms: A comparison of farmer estimate and locomotion scoring. Veterinary Journal 201: 31-38. https://doi.org/10.1016/j.tvjl.2014.05.011

Fadul, M., Bogdahn, C., Alsaaod, M., Hüsler, J., Starke, A., Steiner, A., Hirsbrunner, G., 2017. Prediction of calving time in dairy cattle. Animal Reproduction Science 187: 37-46. https://doi.org/10.1016/j.anireprosci.2017.10.003

Fiedler, A., Bach, K., Pache, S., Jansen, R., Verstijnen, C., Huhne, T., Maierl, J., 2019. Developing a novel concept for preventive hoof maintenance in dairy cows. 20th International Symposium and 12th International Conference on Lameness in Ruminants, Asakusa, Japan March 11-14. Oral presentation, pp. 138-141. Available at: https://tinyurl.com/dhkjbtd8

Food and Agriculture Organization of the United Nations (FAO), 2021. Livestock primary. FAO, Rome, Italy. Available at: http://www.fao.org/faostat/en/#data/QL.

Food and Agriculture Organization of the United Nations (FAO), International Dairy Federation (IDF) and IFCN Dairy Research Network, 2014. World mapping of animal feeding systems in the dairy sector. FAO, Rome, Italy. Available at: http://www.fao.org/3/i3913e/i3913e.pdf.

French, M., 2016. Thermal imaging technology: what can it do for dairy industry? Dairy Reporter. Available at: https://tinyurl.com/97u7nyp7.

Frost, A.R., Schofield, C.P., Beaulah, S.A., Mottram, T.T., J.A., L., Wathes, C.M., 1997. A review of livestock monitoring and the need for integrated systems. Computer and Electronics in Agriculture 17: 139-159. https://doi.org/10.1016/S0168-1699(96)01301-4

Galama, P.J., Ouweltjes, W., Endres, M.I., Sprecher, J.R., Leso, L., Kuipers, A., Klopčič, M., 2020. Symposium review: future of housing for dairy cattle. Journal of Dairy Science 103: 5759-5772. https://doi.org/10.3168/jds.2019-17214

IceRobotics, 2021. Early lameness detection. Available at: https://www.icerobotics.com/cowalert/lameness-alerts/#.

Klootwijk, C.W., De Boer, I.J.M., Van den Pol-van Dasselaar, A., Holshof, G., Fraval, S., Van Middelaar, C.E., 2021. The potential of multispectral images to correct fresh grass allowance for selective grazing. Grassland Science in Europe 26. Available at: https://tinyurl.com/njy5durt

Knight, C.H., 2020. Review: Sensor techniques in ruminants: more than fitness trackers. Animal 14: s187-s195. https://doi.org/10.1017/S1751731119003276

Lokhorst, C., De Mol, R.M., Kamphuis, C., 2019. Invited review: big data in precision dairy farming. Animal 13: 1519-1528. https://doi.org/10.1017/S1751731118003439

Lovarelli, D., Bacenetti, J., Guarino, M., 2020. A review on dairy cattle farming: Is precision livestock farming the compromise for an environmental, economic and social sustainable production? Journal of Cleaner Production 262: 121409. https://doi.org/10.1016/j.jclepro.2020.121409

Lowe, G.L., Sutherland, M.A., Waas, J.R., Cox, N.R., Schaefer, A.L., Stewart, M., 2021. Effect of milk allowance on the suitability of automated behavioural and physiological measures as early disease indicators in calves. Applied Animal Behaviour Science 234: 105202. https://doi.org/10.1016/j.applanim.2020.105202

Lukas, J.M., Reneau, J.K., Linn, J.G., 2008. Water intake and dry matter intake changes as a feeding management tool and indicator of health and estrus status in dairy cows. Journal of Dairy Science 91: 3385-3394. https://doi.org/10.3168/jds.2007-0926

Maltz, E., 2020. Individual dairy cow management: achievements, obstacles and prospects. Journal of Dairy Research 87: 145-157. https://doi.org/10.1017/S0022029920000382

Mazeris, F., 2010. DeLaval herd navigator® proactive herd management. The First North American Conference on Precision Dairy Management. March 2-5, 2010. Toronto, Ontario, Canada. Available at: http://www.precisiondairy.com/proceedings/s1mazeris.pdf.

McManus, C., Tanure, C.B., Peripolli, V., Seixas, L., Fischer, V., Gabbi, A.M., Menegassi, S.R.O., Stumpf, M.T., Kolling, G.J., Dias, E., Costa, J.B.G., 2016. Infrared thermography in animal production: an overview. Computers and Electronics in Agriculture 123: 10-16. https://doi.org/10.1016/j.compag.2016.01.027

McSweeney, D., Coughlan, N.E., Cuthbert, R.N., Halton, P., Ivanov, S., 2019. Micro-sonic sensor technology enables enhanced grass height measurement by a Rising Plate Meter. Information Processing in Agriculture 6: 279-284. https://doi.org/10.1016/j.inpa.2018.08.009

McSweeney, D., O'Brien, B., Coughlan, N.E., Férard, A., Ivanov, S., Halton, P., Umstatter, C., 2020. Virtual fencing without visual cues: design, difficulties of implementation, and associated dairy cow behaviour. Computers and Electronics in Agriculture 176: 105613. https://doi.org/10.1016/j.compag.2020.105613

Meijer, R., Peeters, K., 2010. The use of precision dairy farming in feeding and nutrition. The First North American Conference on Precision Dairy Management. March 2-5, 2010. Toronto, Ontario, Canada. Available at: http://precisiondairy.com/proceedings/s11meijer.pdf.

Mitchell, T., 1997. Machine learning. McGraw-Hill Science, New York, NY, USA.

Moeckel, T., Safari, H., Reddersen, B., Fricke, T., Wachendorf, M., Kumar, L., Mutanga, O., Waser, L.T., Thenkabail, P.S., 2017. Fusion of ultrasonic and spectral sensor data for improving the estimation of biomass in grasslands with heterogeneous sward structure. Remote Sensing 9: 1-14. https://doi.org/10.3390/rs9010098

Molfino, J., Clark, C.E.F., Kerrisk, K.L., García, S.C., 2017. Evaluation of an activity and rumination monitor in dairy cattle grazing two types of forages. Animal Production Science 57: 1557-1562. https://doi.org/10.1071/AN16514

Mottram, T., 2016. Animal board invited review: precision livestock farming for dairy cows with a focus on oestrus detection. Animal 10: 1575-1584 https://doi.org/10.1017/S1751731115002517

Neethirajan, S., 2017. Recent advances in wearable sensors for animal health management. Sensing and Bio-Sensing Research 12: 15-29. https://doi.org/10.1016/j.sbsr.2016.11.004

Neethirajan, S., 2020. The role of sensors, big data and machine learning in modern animal farming. Sensing and Bio-Sensing Research 29: 100367. https://doi.org/10.1016/j.sbsr.2020.100367

Norberg, E., Hogeveen, H., Korsgaard, I.R., Friggens, N.C., Sloth, K.H.M.N., Løvendahl, P., 2004. Electrical conductivity of milk: ability to predict mastitis status. Journal of Dairy Science 87: 1099-1107. https://doi.org/10.3168/jds.S0022-0302(04)73256-7

O'Leary, N.W., Byrne, D.T., O'Connor, A.H., Shalloo, L., 2020. Invited review: cattle lameness detection with accelerometers. Journal of Dairy Science 103: 3895-3911. https://doi.org/10.3168/jds.2019-17123

Oberschätzl-Kopp, R., Haidn, B., Peis, R., Reiter, K., Bernhardt, H., 2016. Effects of an automatic feeding system with dynamic feed delivery times on the behaviour of dairy cows. CIGR-AgEng Conference, June 26-29, 2016. Aarhus, Denmark. Available at: https://www.agroengineering.org/index.php/jae/article/download/869/722/.

Oenema, O., De Klein, C., Alfaro, M., 2014. Intensification of grassland and forage use: driving forces and constraints. Crop and Pasture Science 65: 524-537. https://doi.org/10.1071/CP14001

Roelofs, J., López-Gatius, F., Hunter, R.H.F., Van Eerdenburg, F.J.C.M., Hanzen, C., 2010. When is a cow in estrus? Clinical and practical aspects. Theriogenology 74: 327-344. https://doi.org/10.1016/j.theriogenology.2010.02.016

Roelofs, J.B., Soede, N.M., Kemp, B., 2006a. Insemination strategy based on ovulation prediction in dairy cattle. Vlaams Diergeneeskundig Tijdschrift 75: 70-78.

Roelofs, J.B., Van Eerdenburg, F.J.C.M., Hazeleger, W., Soede, N.M., Kemp, B., 2006b. Relationship between progesterone concentrations in milk and blood and time of ovulation in dairy cattle. Animal Reproduction Science 91: 337-43. https://doi.org/10.1016/j.anireprosci.2005.04.015

Roelofs, J.B., Van Eerdenburg, F.J.C.M., Soede, N.M., Kemp, B., 2005. Pedometer readings for estrous detection and as predictor for time of ovulation in dairy cattle. Theriogenology 64: 1690-703. https://doi.org/10.1016/j.theriogenology.2005.04.004

Roelofs, J.B., Van Erp-van der Kooij, E., 2015. Estrus detection tools and their applicability in cattle : recent and perspectival situation. Animal Reproduction 12: 498-504.

Runhaar, H., Fünfschilling, L., Van den Pol-van Dasselaar, A., Moors, E.H.M., Temmink, R., Hekkert, M., 2020. Endogenous regime change: lessons from transition pathways in Dutch dairy farming. Environmental Innovation and Societal Transitions 36: 137-150. https://doi.org/10.1016/j.eist.2020.06.001

Rutten, C.J., Kamphuis, C., Hogeveen, H., Huijps, K., Nielen, M., Steeneveld, W., 2017. Sensor data on cow activity, rumination, and ear temperature improve prediction of the start of calving in dairy cows. Computers and Electronics in Agriculture 132: 108-118. https://doi.org/10.1016/j.compag.2016.11.009

Rutten, C.J., Steeneveld, W., Inchaisri, C., Hogeveen, H., 2014. An *ex ante* analysis on the use of activity meters for automated estrus detection: to invest or not to invest? Journal of Dairy Science 97: 6869-6887. https://doi.org/10.3168/jds.2014-7948

Rutten, C., Velthuis, A., Steeneveld, W., Hogeveen, H., 2013. Invited review: sensors to support health management on dairy farms. Journal of Dairy Science 96: 1928-1952. https://doi.org/10.3168/jds.2012-6107

Saint-Dizier, M., Chastant-Maillard, S., 2012. Towards an automated detection of oestrus in dairy cattle. Reproduction in Domestic Animals 47: 1056-61. https://doi.org/10.1111/j.1439-0531.2011.01971.x

Saint-Dizier, M., Chastant-Maillard, S., 2015. Methods and on-farm devices to predict calving time in cattle. Veterinary Journal 205: 349-356. https://doi.org/10.1016/j.tvjl.2015.05.006

Scott, P., 2010. Calving Part 1 – the basics. NADIS Animal Health Skills. Available at: https://www.nadis.org.uk/disease-a-z/cattle/calving-module/calving-part-1-the-basics/.

Shalloo, L., Leso, L., Werner, J., Ruelle, E., Geoghegan, A., Delaby, L., 2018. Review: Grass-based dairy systems, data and precision technologies. Animal 10: s262-s271. https://doi.org/10.1017/S175173111800246X

Slob, N., Catal, C., Kassahun, A., 2021. Application of machine learning to improve dairy farm management: A systematic literature review. Preventive Veterinary Medicine 187: 105237. https://doi.org/10.1016/j.prevetmed.2020.105237

SmaXtec, 2021. Reproduction: heat detection and calving prediction. SmaXtec, Graz, Austria. Available at: https://smaxtec.com/en/modules/reproduction/.

Stichting Kwaliteitszorg Onderhoud Melkinstallaties, 2021. Statistiek. Stichting KOM, Zutphen, the Netherlands. Available at: https://stichtingkom.nl/index.php/stichting_kom/category/statistiek

Strucken, E.M., Laurenson, Y.C.S.M., Brockmann, G.A., 2015. Go with the flow-biology and genetics of the lactation cycle. Frontiers in Genetics 6: 118. https://doi.org/10.3389/fgene.2015.00118

Swartz, T.H., Findlay, A.N., Petersson-Wolfe, C.S., 2017. Short communication: automated detection of behavioral changes from respiratory disease in pre-weaned calves. Journal of Dairy Science 100: 9273-9278. https://doi.org/10.3168/jds.2016-12280

Tanaka, T., Arai, M., Ohtani, S., Uemura, S., Kuroiwa, T., Kim, S., Kamomae, H., 2008. Influence of parity on follicular dynamics and resumption of ovarian cycle in postpartum dairy cows. Animal Reproduction Science 108: 134-143. https://doi.org/10.1016/j.anireprosci.2007.07.013

Thorup, V.M., Munksgaard, L., Robert, P.-E., Erhard, H.W., Thomsen, P.T., Friggens, N.C., 2015. Lameness detection via leg-mounted accelerometers on dairy cows on four commercial farms. Animal 9: 1704-1712. https://doi.org/10.1017/S1751731115000890

Titler, M., Maquivar, M.G., Bas, S., Rajala-Schultz, P.J., Gordon, E., McCullough, K., Federico, P., Schuenemann, G.M., 2015. Prediction of parturition in Holstein dairy cattle using electronic data loggers. Journal of Dairy Science 98: 5304-5312. https://doi.org/10.3168/jds.2014-9223

Van De Gucht, T., Saeys, W., Van Weyenberg, S., Lauwers, L., Mertens, K., Vandaele, L., Vangeyte, J., Van Nuffel, A., 2017. Automatic cow lameness detection with a pressure mat: effects of mat length and sensor resolution. Computers and Electronics in Agriculture 134: 172-180. https://doi.org/10.1016/j.compag.2017.01.011

Van de Haar, M., Armentano, L., Weigel, K., Spurlock, D., Tempelman, R., Veerkamp, R., 2016. Harnessing the genetics of the modern dairy cow to continue improvements in feed efficiency. Journal of Dairy Science 99: 4941-4954. https://doi.org/10.3168/jds.2015-10352

Van den Pol-van Dasselaar, A., Becker, T., Botana Fernández, A., Peratoner, G., 2021. Societal and economic options to support grassland based dairy production in Europe. Irish Journal of Agricultural and Food Research. https://doi.org/10.15212/ijafr-2020-0128

Van den Pol-van Dasselaar, A., Hennessy, D., Isselstein, J., 2020. Grazing of dairy cows in Europe-an in-depth analysis based on the perception of grassland experts. Sustainability 12: 1098. https://doi.org/10.3390/su12031098

Van Erp-van der Kooij, E., Rutter, S.M., 2020. Using precision farming to improve animal welfare. CAB Reviews 15. https://doi.org/10.1079/pavsnnr202015051

Van Nuffel, A., Saeys, W., Sonck, B., Vangeyte, J., Mertens, K.C., De Ketelaere, B., Van Weyenberg, S., 2015. Variables of gait inconsistency outperform basic gait variables in detecting mildly lame cows. Livestock Scienceivestock Science 177: 125-131. https://doi.org/10.1016/j.livsci.2015.04.008

Vázquez-Diosdado, J.A., Miguel-Pacheco, G.G., Plant, B., Dottorini, T., Green, M., Kaler, J., 2019. Developing and evaluating threshold-based algorithms to detect drinking behavior in dairy cows using reticulorumen temperature. Journal of Dairy Science 102: 10471-10482. https://doi.org/10.3168/jds.2019-16442

Voß, A.L., Fischer-Tenhagen, C., Bartel, A., Heuwieser, W., 2021. Sensitivity and specificity of a tail-activity measuring device for calving prediction in dairy cattle. Journal of Dairy Science 104: 3353-3363. https://doi.org/10.3168/jds.2020-19277

Wang, J., Bell, M., Liu, X., Liu, G., 2020. Machine-learning techniques can enhance dairy cow estrus detection using location and acceleration data. Animals 10: 1160. https://doi.org/10.3390/ani10071160

Werner, J., Leso, L., Umstatter, C., Niederhauser, J., Kennedy, E., Geoghegan, A., Shalloo, L., Schick, M., O'Brien, B., 2018. Evaluation of the RumiWatchSystem for measuring grazing behaviour of cows. Journal of Neuroscience Methods 300: 138-146. https://doi.org/10.1016/j.jneumeth.2017.08.022

West, J.W., 2003. Effects of heat-stress on production in dairy cattle. Journal of Dairy Science 86: 2131-2144. https://doi.org/10.3168/jds.S0022-0302(03)73803-X

Wilkinson, J.M., Lee, M.R.F., Rivero, M.J., Chamberlain, A.T., 2020. Some challenges and opportunities for grazing dairy cows on temperate pastures. Grass and Forage Science 75: 1-17. https://doi.org/10.1111/gfs.12458

Williams, M., Mac Parthaláin, N., Brewer, P., James, W., Rose, M., 2016. A novel behavioral model of the pasture-based dairy cow from GPS data using data mining and machine learning techniques. Journal of Dairy Science 99: 2063-2075. https://doi.org/10.3168/jds.2015-10254

Wu, Y., Neerchal, N., Dyer, R., Tasch, U., Rajkondawar, P., 2011. Modeling bovine lameness with limb movement variables. Journal of Biomedical Science and Engineering 4: 419-425. https://doi.org/10.4236/jbise.2011.46053

3. Precision beef and sheep farming

F. Maroto-Molina[*] and D.C. Pérez-Marín

Department of Animal Production, School of Agricultural and Forestry Engineering, University of Cordoba, Campus de Rabanales, Ctra. Madrid-Cádiz, km 396, 14071 Córdoba, Spain; fmaroto@uco.es

Highlights

- Beef cattle and sheep are important livestock sectors in Europe, especially in vulnerable areas, where they provide high-quality products and numerous ecosystem services.

- There are few PLF solutions specifically designed for beef cattle and sheep, but they benefit from technology developments for other sectors, such as dairy farming.

- PLF solutions can be classified as wearable and non-wearable, according to their physical relationship to the animal. Wearable devices are normally used with large and high-value animals, while non-wearable options are chosen when technologies are large, complex, or expensive.

- The most common PLF solutions for beef cattle and sheep are focused on location, especially in extensive farming systems, and weight monitoring, because producing meat is the primary aim.

- Radio Frequency Identification is mandatory for sheep, posing a great opportunity to make them readable by different PLF tools.

- PLF solutions to monitor the fields, based on proximal or remote sensors, are promising tools to increase production efficiency and system sustainability.

3.1 Background of beef cattle farming in Europe

The European cattle sector consists of the bovine meat sub-sector, which produces and processes beef and veal, and the dairy sector, which produces and processes cow milk. It is important to clarify that meat production comes both from beef cows (also known as suckler cows) and dairy cows, being the production systems of both subsectors quite different. In this chapter, the focus is on beef cattle production systems, as dairy farming systems are addressed in Chapter 2.

Nowadays, the European Union (EU) is the world's third producer of beef with 7,844 million tons of meat in 2019 (EUROSTAT, 2021), behind the United States of America (USA) and Brazil. In 2019, there were over 86.5 million bovine animals in the EU. This figure comprises all types of bovine animals, i.e. beef and dairy herds, and animals of different ages. Specifically, the number of cows was 34.8 million, being 22.6 million dairy cows and the rest suckler cows (EUROSTAT, 2021). The EU cattle sector is biased toward dairy production. Therefore, beef production is divided into two main types or subsectors: dairy-beef production, which is a primary driver of the EU beef production sector, and beef breed production, focused on cow-calf systems. The EU dairy cattle production is concentrated mainly in six countries, accounting for almost 70% of total heads (Germany, France, Poland, Italy, United Kingdom (UK), and the Netherlands). On the other hand, France, Spain, UK, and Ireland hold over 70% of the EU suckler cows, which are normally used to valorise less-favoured areas (Lherm *et al.*, 2017). The highest number of fattening farms corresponds to Ireland, followed by Spain, the regions in or around the Alps, Eastern Poland, and Slovenia. Fattening farms in those regions have a small average size, while the largest farms are in Germany and the Benelux (Hocquette *et al.*, 2018).

Beef consumption in Europe is about 16 kg per capita (20% of total meat consumption), far below other countries, such as Argentina, Brazil, or the USA, with annual consumptions per capita between 37 and 54 kg (Ihle *et al.*, 2017). Although a small reduction in domestic beef demand has been observed in the last years, the overall beef production in the EU is expected to maintain stable, due to the increasing demand of developing countries, such as China.

The heterogeneity of the EU cattle sector at the regional level is very pronounced. Beef cattle farms under extensive, semi-extensive or intensive regimes can be found, depending on the region and the production phase. Beef farming practices and farm structures vary largely throughout the EU because of climatic factors and the large variety of ecological zones in Europe. Beef production systems in the EU also differ regarding the age and weight at which animals are slaughtered, the feeding regime and the type of accommodation. Nevertheless, in general, two production phases can be distinguished:

1. Breeding, also referred to as cow-calf systems, which are dedicated to the production of calves from suckler cows. It also includes heifer rearing for the replacement of the herd.
2. Finishing, which refers to the fattening of calves produced in breeding farms and dairy farms, prior to the slaughterhouse.

Individual farms can be dedicated to breeding, finishing or both (complete cycle farms), although specialisation is the most common option in the EU beef cattle sector, i.e. production phases tend to be separated in different farms and even in different regions.

Normally, cow-calf farms and complete cycle farms are pasture-based systems, where animals have permanent access to pasture. In some warm regions, mostly in Mediterranean countries, where grass production can be 5 to 10 times lower than in Northern Europe, animals require significant feeding supplementation during several months of the year. It is worth highlighting the case of Southwestern Spain, where beef production is linked to a very specific agroforestry system, named 'dehesa', where cattle, sheep and Iberian pigs are raised together under predominantly oak trees. Agroforestry refers to the land management systems combining the growing of trees, crops and livestock in the same place (Allen *et al.*, 2011). In drylands, agroforestry systems play crucial economic, social, and environmental roles, reaching high levels of sustainability.

During the breeding phase, calves from suckler cows remain with their mothers for a four-to-eight-month period before they are weaned. This is a very important difference with calves coming from dairy farms, which are separated from their mothers when they are one or two days old and artificially reared on milk or milk replacer plus solid feed for a two-to-three-month period. The life cycle of animals in cow-calf farms is represented in Figure 3.1.

In breeding farms, reproductive efficiency is essential to obtain a satisfactory economic return for the farmer. Thus, some indicators, such as the percentage of pregnant cows, the proportion of born alive and weaned calves, and the calving

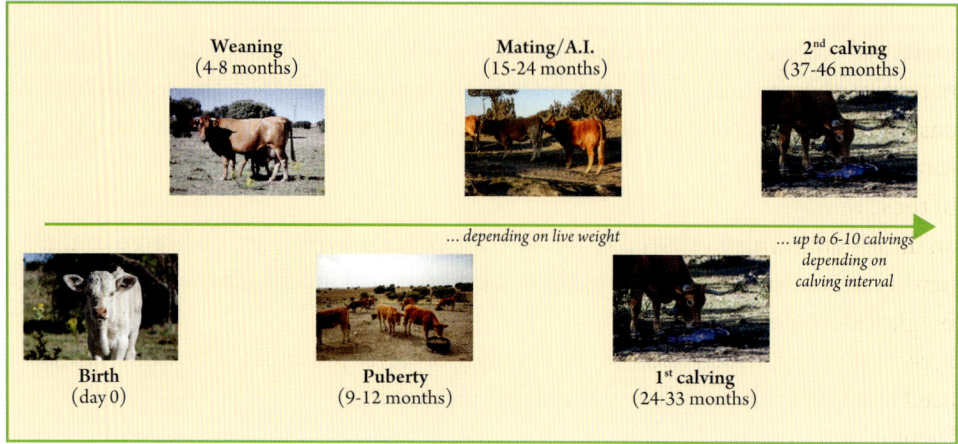

Figure 3.1. Timeline for beef cattle from birth to first and second time calving (photos: Rubén Blanco Carrera, farm La Blaquería).

interval are used. One of the most important management decisions is to establish controlled calving seasons. Despite the widespread adoption of artificial insemination (AI) worldwide, natural service is the most frequent breeding method in beef cattle farms, mainly due to the management constraints associated with the extensive or semi-extensive conditions under which these animals are reared. Besides, extensive production systems make it difficult to gather information on animal behaviour, feeding status or health that allow the making of timely decisions. Closer monitoring of animals and timely interventions could greatly reduce the number of cows failing to become pregnant and the number of calves lost during the calving season. This is an area of research for many Precision Livestock Farming (PLF) applications, as explained later in this chapter.

Feeding optimisation, which is needed to maintain animal health and productivity, is another field of interest for beef cattle farming. In pasture-based systems, most feed comes from grass, so monitoring grass growth and consumption is as crucial as monitoring cows. Grass production and utilisation are below optimum in most farms. Grass underutilisation, as well as overgrazing, should be avoided, improving grazing efficiency and saving significant costs for the farmer through the minimisation of supplementary feed. Stocking rate (number of livestock units per area unit) is the indicator most widely used in this field and fences are the main management tool.

Regarding the fattening phase, production systems may be divided into extensive (grass-based), or intensive (indoor), the latter being the most common in Europe.

Different production systems are associated with different feeding regimes: grazing natural pastures (daily gain of 0.3 kg of live weight (LW) a day), grazing seeded pastures (daily gain of 0.6 to 1.0 kg LW a day) or fed concentrates plus different forages at feedlots (daily gain over 1.2 kg LW a day), although some combinations of these options also exist. The diverse climatic conditions of the EU countries have an important influence on the type of fattening system and the breeds used. In beef herds, it is very usual to rear different breeds (Limousine, Charolais, Aberdeen Angus, Belgian Blue, Retinta, Chianina … among a list of more than 200 specialised and local breeds), in contrast to dairy farms, where Friesian and Holstein breeds are extensively used in all EU countries. The use of crossbred animals is also widespread in beef cattle farming, enabling farmers to obtain better growth rates and carcass yields.

The fattening phase usually includes two steps: (1) the backgrounding period, which comprises the first 60 days after arrival at the feedlot, when the animals adapt themselves to the new feeding and housing system, and (2) finishing stage, with a duration of three to five months (if the fattening phase is extensive this period will be much longer, up to 12 to 14 months), when animals are fed mainly concentrates complemented with some forages, in most cases just cereal straw or silage. Average slaughter weights are 450 to 600 kg for heifers and steers coming from beef herds, and 350 to 450 kg for animals coming from dairy herds (dairy animals have an earlier deposition of fat).

During the fattening phase, live weight and daily gain are key indicators, as well as feed intake. The continuous monitoring of these indicators, together with carcass conformation and fat score, could allow the calculation of feed efficiency and the identification of the optimal slaughter endpoint from an economic point-of-view, both having an important impact on the profitability of fattening farms. However, these data are normally not gathered in fattening farms, due to cost and management constraints.

3.2 Background of sheep farming in Europe

Sheep meat and milk production is an important industry in Europe, especially in the UK, Spain, Ireland, France, Italy, Greece, and Romania. There are 85 million sheep in around 800,000 farms in the EU (EUROSTAT, 2021), often located in economically vulnerable areas, such as mountainous and dry regions. Despite this large number of sheep, Europe is only 85% self-sufficient in sheep meat, being the world's second-largest importer of sheep meat. New Zealand is the EU's biggest supplier, accounting for about 80% of imports, followed by Australia. The EU exports around 10% of its total production. Live sheep are mostly traded to the Middle East and North Africa, while sheep meat is predominantly shipped to the Far East. The sheep industry is declining in Europe (except for the UK) and during the past 40 years, sheep production has decreased by up to 40% depending on the country and the farming system. This situation is mainly attributed to the low profitability of sheep farming in comparison to other agricultural activities and the poor structure of the sheep industry (small farm size, the average age of sheep farmers, etc.). Even in ecologically vulnerable areas, sheep are being substituted by beef cattle (less suitable to these areas) because of the higher labour needs of sheep (Maroto-Molina *et al.*, 2018).

An important attribute of ovine is that they can live and produce on land unfavourable to other forms of agriculture. Many sheep breeds are adapted to survive on extensive, unimproved natural pastures and difficult climate conditions. Sheep are also very valuable for small farms, where large ruminants may have problems due to limited land and resources, and when small quantities of meat are required for local consumption in remote communities (Morris, 2017). As sheep farming is usually an activity associated with areas where there are no economic alternatives, it has a primary role in structuring the territory, preserving the environment, and generating employment in rural areas. In recent years, its ability to adapt to new market conditions has become evident, increasing trade with third countries, as well as promoting and enhancing the quality of sheep products.

Normally, sheep farms produce more than one product, including meat, milk, wool, and skins. Around the world, wool production is declining, displaced by synthetic fibres. Only Merino wool has an acceptable demand. Meat is the most relevant sheep product from an economic point of view, followed by milk. Within the EU, the UK and Spain are the two main producers of sheep meat. The type of meat production differs among regions. Heavy lambs (over 13 kg LW) are mostly produced in Ireland and Romania, light lambs are reared in southern countries, such as Italy and Greece, whereas UK, Spain and France have both types of productions. At the European level, there is a decline in sheep meat consumption, which can be attributed to the cooking difficulties associated with this type of meat (cooking time), the lack of appeal of sheep meat, especially among young consumers, and the shift of consumption trends from red meat to pork and poultry. Nevertheless, there are some exceptions to this general trend, such as Germany, where lamb demand is increasing due to changing demographics.

The EU leads the production of sheep milk, linked to high-quality cheese production, especially in southern and eastern countries, such as Spain and Romania. It is worth highlighting the case of France, where dairy sheep are reared under intensive systems, using highly specialised breeds. Some of these cheese productions are protected under quality labels, as a means to increase their added value and to attract consumers.

The type of sheep production system depends on the environment, the degree of control, the management programme, and, especially, the main product obtained from the animals. Farming systems include extensive farms based on pasture, as well as intensive indoor productions taking advantage of modern technologies and management options. In general, extensive regimes are linked to meat production (or dual-purpose), while intensive systems are more associated with dairy sheep farming.

Traditionally, pastoralism has been the main sheep production system, especially in semiarid rangelands. These production systems were characterised by the movement of animals searching for new pastures, from highly nomadic to transhumance (FAO, 2001). Currently, nomadism does not exist in the EU, while transhumance is disappearing, due to labour costs and the gradual expansion of arable land (Morris, 2017). Nowadays, sheep produced in extensive farms normally graze within enclosed fenced systems.

In most sheep herds, nutritional and reproductive management levels are low to medium depending on the profitability of each farming system. Regarding reproduction, it is important to consider that sheep are seasonal polyoestrous animals, i.e. their oestrus cycles are affected by season, particularly by the photoperiod in short days. This means that they start oestrus cycles when the length of the day is decreasing and return to anoestrus during the spring. Thus, most breeding occurs naturally between October and November (lambing in March and April), unless specific reproductive techniques, such as hormone treatments or male effect are applied. Adequate programming of mating and lambing seasons is essential for the economic success of a sheep farm. Normally, natural service is preferred to artificial insemination.

In the case of meat-oriented sheep farms, lambs stay with their mothers until weaning, which may vary from four weeks to four months (the life cycle of sheep is shown in Figure 3.2). However, when milk is the primary production, efforts are directed to increase commercial milk yield, and lamb meat is considered a secondary product.

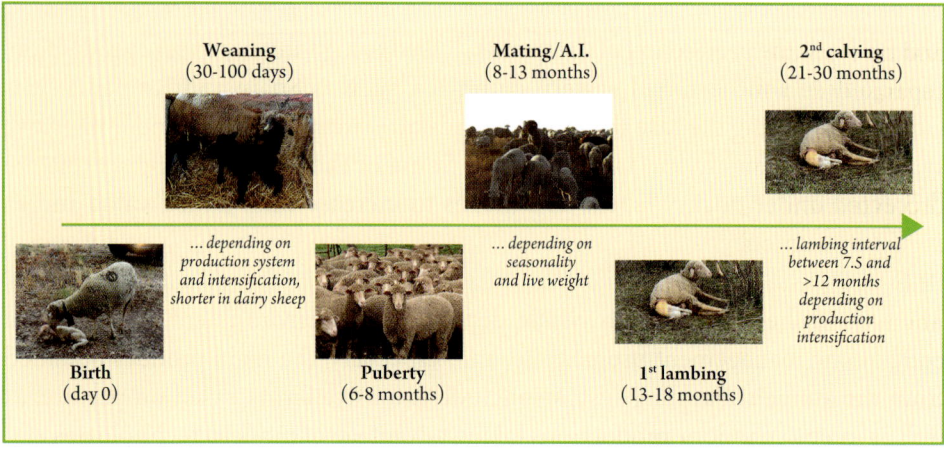

Figure 3.2. Timeline for sheep from birth to first and second time lambing (photos: Felipe Molina, farm Las Albaidas).

To increase milk yield, weaning age is reduced and sometimes artificial rearing with milk replacers is used. For the fattening phase, lambs may stay in the farm for grazing (until they are five to eight months old) or be moved to feedlots, where they are fed on concentrate, often for a short period and after early weaning. The fattening phase can oscillate between two and six months depending on the type of meat to be commercialised, i.e. light, or heavy carcasses. Grass-fed lamb production is very challenging due to several constraints, including seasonality of pasture availability and climate variability, which make it very difficult to maintain product consistency.

In extensive sheep farming, the two main indicators of production efficiency are stocking rate and the number of lambs weaned per ewe, which determine the quantity of lamb produced per hectare. Key factors affecting production efficiency are ewe genotype, farm management and nutrition. It is important to highlight the large number of sheep breeds available in Europe, and the opportunities for the local breeds to perform well due to their high adaptation to specific environments and their association with high-quality products. Purebred animals are normally used for breeding. Crossbred animals are used for fattening to improve feed conversion rates and carcass yield.

In intensive production systems, related to dairy farms, ewes are usually milked twice a day during a lactation period of three to six months, yielding between 400 and 600 litres per lactation period. Milk production peaks at four to seven weeks postpartum and then gradually decreases. Lambs stay with their mothers for four weeks or are separated at birth to be reared artificially.

3.3 Application of PLF in beef cattle and sheep

PLF solutions can be classified as wearable and non-wearable, according to their physical relationship to the animal. Wearable devices are normally used with large and high-value animals, such as dairy cows. They are also the preferred option when cheap technologies, e.g. accelerometers or thermistors, are used. On the other hand, non-wearable options are chosen when technologies are large, complex, or expensive (Caja *et al.*, 2020).

Wearable devices have certain requirements conditioning their use in beef cattle and sheep, especially in the case of extensive farming systems:

1. They need to be wireless, i.e. to have the ability to communicate the gathered data to the processing unit, normally a cloud server, where these data are analysed to support decision-making. Inside a barn, several options are available to establish wireless networks, e.g. Wi-Fi or Bluetooth. In the case of extensive farming systems, animals can be dispersed in very large areas, and wireless communication options are much more limited. The Global System for Mobile (GSM) communications has been the most widely used service to transmit data in open spaces (Tomkiewicz *et al.*, 2010). Nevertheless, although GSM services are available in most EU cities and towns, there are still vast areas without them, mostly rural areas, where beef cattle and sheep herds are located. In those cases, Low-Power-Wide-Area (LPWA) networks, such as Sigfox or LoRa are the main alternative. They provide long-range communication up to 10 to 40 km. LPWA networks are highly energy-efficient and inexpensive, but the size of data that can be transferred through them is very limited (normally few bytes per message and few messages per device and day). This fact highly influences the design of PLF applications (Maroto-Molina *et al.*, 2019).
2. They need to be small and compact, as animals will carry them on the neck, the ear, etc. This is especially important in terms of battery size. To avoid PLF devices from being excessively heavy, a balance between battery duration and sampling interval must be achieved. A long sampling interval using a sleep/burst mode may not provide the level of data resolution

necessary to classify specific behaviours (Theurer *et al.*, 2013). It is worth highlighting that there are not many PLF applications specifically designed for beef cattle and sheep. In most cases, PLF technologies developed for dairy cows are adapted to beef cattle, but to be fitted to small ruminants they would need a reduction from 1:2 to 1:5 scale, because of the large size and weight of cattle devices. This reduction entails several technological challenges, as there is a risk of a reduced communication range (smaller antenna) and life span (lower weight of batteries).
3. They need to be robust and resistant to harsh environments to avoid breakages and failures. This is a commodity feature for PLF wearable devices, but it is especially necessary in the case of animals reared on grazing systems. Under extensive conditions, PLF devices may be exposed to rain, large variations in temperature, interactions between animals and the different elements of the landscape (e.g. trees and rocks) and even fights among animals. Additionally, as in many extensive systems animals are not frequently managed, wearable devices may be fitted to animals for long periods, limiting the time available for adjustments and reparations.

Non-wearable devices also pose some challenges in extensive production systems, such as power supply or connectivity, but their adaptation, e.g. by using solar panels, is easier than in the case of wearable solutions.

In the next sub-sections, the main PLF applications available for beef cattle and sheep are described.

3.3.1 Radio frequency identification

The radio frequency spectrum is organised into bands, as agreed by the International Telecommunication Union (ITU). Greater frequency (lower wavelength) requires the use of greater energy supply to produce the electromagnetic waves, provides greater penetration, and hence is associated with greater potential risk. Moreover, penetration depends on the absorption and scattering properties of the material (e.g. the penetration of infrared into the skin increases with its water content). Electromagnetic waves may be considered damaging radiations for living organisms when frequency surpasses infrared, which includes visible and ultraviolet light (skin damages), radar (burns), X-rays (cancer), and y-rays (lethal for microorganisms) (Caja *et al.*, 2020). It is important to note that distance to the source and body size affect the total amount of energy received, which should be minimised to maximise safety.

Animal Radio Frequency Identification (RFID) uses the low frequency (LF) band, which is very safe. RFID requires LF waves because they can pass through water and animal tissues, although they are only readable at short distances and with large antennas. This makes possible the use of RFID as part of a bolus located in the rumen. In contrast, high frequency (HF) and ultra-high frequency (UHF) waves are poorly propagated through water or animal tissues, but they can reach long distances and be read with small antennas. Consequently, UHF is used for long-range communication, e.g. for data transmission from the sensor to the server.

The use of RFID is mandatory for small ruminants. This is a consequence of the bovine spongiform encephalopathy epidemic of the 1990s and the recognition of related problems (Scrapie) in small ruminants. The use of a tamperproof and effective tagging system was required to implement reliable health and tracing controls in small ruminants, which was non-existent at that time. The EU made compulsory the use of electronic identification for most breeding stock (older than 6 months). Despite regional differences in acceptance and speed of implementation, the use of RFID was finally fully adopted in the whole EU in 2010 (Caja *et al.*, 2020).

RFID transponders are passive devices (without battery), which respond with a coded signal when they are activated by the electromagnetic pulses produced by a transceiver (reader). There are several ways to attach them to animals. Glass encapsulated injectable transponders were the first wearable device to be used for the electronic identification of ruminants, mainly in sheep. Injectables are currently not recommended or even banned under commercial farm conditions, because of consumer concerns resulting from the difficult or uncertain retrieval of the transponder at slaughter. Ear tags are the preferred option for RFID transponders in the case of cattle and pigs because of their easy insertion and reading. However, because of ear features, tissue consistency and propensity to chewing, ear tag dimensions in small ruminants should be smaller than in cattle or pigs, aiming to prevent ear tearing and ear tag losses. Considering that, rumen boluses are normally the chosen option for the electronic identification of sheep because of their high retention and tamperproof features. Because of their location in the rumen, high-performance LF devices are needed. The ideal bolus for oral application in small ruminants should be less than 20×100 mm and 70 g, with a specific gravity greater than 2.5 to warrant its retention in the rumen (Caja *et al.*, 2020).

It is worth highlighting that RFID is not a sensing technology, but it makes animals readable by different PLF applications described in the next sub-sections, such as weighing systems.

3.3.2 Global positioning systems

The Global Positioning System (GPS) is a set of 31 navigation satellites circling the Earth. The USA, which developed and operates GPS, and Russia, which developed a similar system known as GLONASS, have offered free use of their respective systems to the international community. The International Civil Aviation Organisation (ICAO) has accepted GPS and GLONASS as the core for an international civil satellite navigation capability known as the Global Navigation Satellite System (GNSS). The basic GPS service provides users with approximately 7.8-meter accuracy, 95% of the time, anywhere on or near the surface of the Earth. To accomplish this, each of the 31 satellites emits signals to receivers that determine their location by computing the difference between the time that a signal is sent and the time it is received. GPS satellites carry atomic clocks that provide extremely accurate time. The time information is placed in the codes broadcast by the satellite so that a receiver can continuously determine the time the signal was broadcast. The signal contains data that a receiver uses to compute the locations of the satellites and to make other adjustments needed for accurate positioning. With information about the ranges to three satellites and the location of these satellites when the signal was sent, the receiver can compute its own three-dimensional position. An atomic clock synchronised to GPS is required to compute ranges from these three signals. However, by using a measurement from a fourth satellite, the receiver avoids the need for an atomic clock. Thus, the receiver uses four satellites to compute latitude, longitude, altitude, and time (FAA, 2021).

Several commercial GPS-based solutions are currently used for the location of grazing animals. Remote location is of special interest for cows and sheep grazing in large areas, saving time and other costs (e.g. fuel) for the farmer. At the same time, location data can be used to set geofencing and theft alarms (i.e. the farmer is informed when the location of the animal correspond to a point outside a previously defined perimeter). Most commercial systems have user-friendly mobile apps where users can check the position of monitored animals on an interface based on Google Maps or a similar map service.

Depending on the sampling interval, animal location data can also be used to build trajectories, as a means to characterise grazing behaviour. Deviations from normal behaviour may be indicative of calving/lambing (Fogarty *et al.*, 2020), oestrus or predator attacks. Nevertheless, these alarm systems are still not fully implemented in commercial devices.

Virtual fencing is one of the most desirable PLF tools for extensive farms. This term refers to a structure serving as an enclosure or boundary for animals without the existence of a physical barrier. This system could lower the cost and landscape impacts associated with fencing and reduce labour needs (Umstätter, 2011), the latter being especially important in the case of sheep farming. Early forms of virtual fencing relied on the electromagnetic coupling of a wearable collar and an induction cable, either placed on the ground (without poles) or buried, to delimit the paddock (Anderson, 2007). Nowadays, commercial virtual fencing solutions mostly rely on GPS technology. Virtual fencing pairs an alert signal, that can be audio or vibration, with a negative stimulus, in most cases an electric pulse of low intensity. Normally, these systems start emitting the alert signal when the animal approaches the virtual boundary, allowing it to turn around. Several warning zones are defined with different intensity of the alert signal, but if the animal remains in the warning zone, an electric shock is administered. Through the pairing of the alert and the negative stimulus, the animal learns to avoid the boundaries by responding to the alert signal alone. However, there is substantial variability in the learning rate, which poses some ethical concerns (Marini *et al.*, 2018).

GPS technology presents several challenges limiting its widespread adoption by beef cattle and sheep farmers, including power efficiency, data transmission, cost, size, and accuracy. As GPS are active devices that work in the UHF band, they have large energy needs. Reducing sampling interval means more location data and additional possibilities for data analysis, but, at the same time, it shortens battery life, which needs to be at least several months in extensive production systems, where animals are not accessible for battery change most of the time. A compromise solution optimising the relationship between sampling interval and battery size needs to be found in each case. Most current commercial devices provide location data for each period of 30 minutes to two hours. Data transmission, which is needed for decision support, can also be an important challenge, as data coverage is often not available in mountainous and rural areas. Satellite Internet access can solve this problem, but it currently has a prohibitive cost for most PLF applications. Device cost itself can be another key limitation, especially for sheep farmers. To reduce costs, farmers could only monitor a part of the grazing herd, assuming a gregarious behaviour of non-monitored animals. However, this is incompatible with some GPS applications, such as virtual fencing. A different approach to cost reduction has been tested by Maroto-Molina *et al.* (2019), consisting of some animals in the herd being fitted with GPS collars and the rest with low-cost Bluetooth Low Energy (BLE) ear tags. These tags have a low energy demand (small-sized) and can communicate with GPS collars placed at a short distance, which allows gathering data about tag location (relative to the location of the closest GPS collars). The accuracy of tag location data is low, but it could be sufficient for many utilities. Finally, the accuracy of data provided by GPS tools can also be challenging for the development of some applications. Most PLF devices provide an accuracy between 10 and 30 m, which can be affected by the surrounding landscape, i.e. topography and canopy cover.

Neck collars are the preferred option for GPS-based applications, allowing a larger battery size in comparison to ear tags. However, GPS ear tags are also commercially available, mostly used in beef feedlots for early disease detection based on the modelling of animal behaviour. These devices have a longer sampling interval and a shorter life span (Schleppe *et al.*, 2010). PLF applications based on GPS commonly integrate other sensors aimed to characterise animal behaviour, especially accelerometers.

3.3.3 Accelerometers

Accelerometers are sensors that measure proper acceleration, which is the acceleration (change of rate of velocity) of a body in its instantaneous rest frame. This is different from coordinate acceleration, which is acceleration in a fixed coordinated system. For example, an accelerometer at rest on the Earth's surface will measure gravity acceleration, by definition, straight upwards and 9.81 m/s^2. In contrast, an accelerometer in free fall will measure zero.

Accelerometers are one of the most used PLF sensors in all livestock species. The most useful data come from three-axial accelerometers, which allow the measurement of total activity (motion index) and changes of position in a three-dimensional frame. They can register movement patterns linked to behaviours such as resting, walking, ruminating, grazing, etc. (Fogarty *et al.*, 2018). Deviations from normal behaviour are used to detect calving, oestrus (Valenza *et al.*, 2012), lameness (O'Leary *et al.*, 2020), disease, or heat stress (Islam *et al.*, 2020).

Accelerometers can be fitted to very different devices: collars, ear tags, leg bracelets, halters, ruminal boluses, etc. The position of the accelerometer in relation to the animal will depend on the use of acceleration data. For example, if the goal is to quantify grazing behaviour, the sensor will preferably be placed on a neck collar or a halter. For lameness detection, a leg bracelet will be the preferred option. Ear tag accelerometers have been used to quantify rumination time in beef cattle (Wolfger et al., 2015). Accelerometers inserted in rumen boluses have been used to monitor rumination and gastrointestinal motility (Hamilton et al., 2019).

Even though accelerometers are a mature technology, data interpretation is still a field of research. Acceleration data have some limitations to infer animal position or behaviour, which could be partly overcome by integrating accelerometers, gyroscopes, and magnetometers, together named as inertial measurement units (IMU). Preliminary experiments have demonstrated the potential use of IMU to detect complex behaviours, such as grass intake (Andriamandroso et al., 2017).

Moreover, their integration with batteries and data communication systems into commercial solutions is still under development in many cases. Several accelerometer-based commercial solutions have been designed for dairy cows, but the number of tools specifically designed for beef cattle and sheep is very limited. In this case, developers have focused their effort mostly on feedlot cattle, developing ear tag systems aimed at the identification of sick animals. It is worth mentioning that in a feedlot context, data transmission requirements may be limited to several hundreds of meters, which facilitates the design of small devices.

3.3.4 Pedometers

A pedometer is a device, usually electronic or electromechanical, that counts steps by detecting the motion of legs or hips. Pedometers are routinely used for oestrus detection in dairy cattle farms. Some companies offer pedometers for oestrus detection also in beef cows, but these systems are only valid for semi-extensive production systems since they need to connect to antennas with a range coverage of about 700 m. In the case of sheep, pedometers have been used only for research purposes. Powell (1968) was the first to use a pedometer in sheep to compute the distance walked by animals. Results of this experiment were unreliable due to the heterogeneity of walking patterns of sheep, also influenced by climate conditions. For example, ewes showed excessive shaking on rainy days.

3.3.5 Proximity loggers

Proximity loggers are devices that both transmit and receive radio signals. They are connected through the emission of an identified pulse and the reception of other nearby signals. Proximity loggers record the radio signal strength as a 'received signal strength indicator' (RSSI) value, which can be calibrated against distance under controlled conditions (i.e. check the effect of external factors over the attenuation of radio waves). Subsequently, the distance obtained may correspond to contact between two individuals according to a specific definition of it. The reception between devices is far from perfect since some of the potential contacts are not recorded owing to the internal malfunction of the devices and/or the effect of external factors (Triguero-Ocaña et al., 2019).

Proximity loggers are employed to record proximity events between animals or between animals and tagged points and to estimate their frequency. In livestock science, they have been used to study the social interactions between animals in a herd (Patison et al., 2017), or to match lambs/calves to their mothers, both with breeding and management purposes (Swain and Bishop-Hurley, 2007). Nevertheless, the automatic detection of mating events is probably their main PLF application, both for beef cattle (O'Neill et al., 2014) and sheep.

In the case of sheep, a semi-commercial system (Alpha-D) was developed by Alhamada et al. (2017). It consists of a harness for the ram with an active reader and transmitter which detects the ewes' electronic identifier (RFID), glued to its tail, and transmit the data to a centralised computer. When the ram mounts, a pressure sensor triggers the reading of the transponder attached to the tail of the female. At each reading, the identification of the female and the exact date and time are recorded. Data gathered by Alpha-D can be used to monitor the frequency of mating, true and false coverings, and the number of ewes covered. Thus, it is useful for both oestrus detection and the evaluation of ram mating behaviour. Strictly speaking, Alpha-D is not a proximity-logger application, as female devices do not record contacts, but it was included in this sub-section because the basic principle of this PLF tool is the same as for proximity loggers.

3.3.6 Microphones

Sound analysis is a promising tool for many PLF applications. It has been mostly used in pig and poultry farming. Regarding ruminants, a promising application is the estimation of grass intake using acoustics to identify chewing and biting, both in beef cattle (Galli *et al.*, 2018) and sheep (Sheng *et al.*, 2020). Intake is a key indicator of performance and efficiency, but it is very difficult to measure at the individual level, especially in grazing systems. Unfortunately, this PLF application is not commercially available at present.

The use of microphones and sound analysis has also been proposed as a tool to evaluate welfare status through the study of animal vocalisations (De la Torre *et al.*, 2015). Vocalisations play key roles in a wide range of communication contexts, e.g. for individual recognition and to help coordinate social behaviours. They can also be a signal of pain or fear.

3.3.7 Thermistors

Temperature sensors are often thermistors, which are resistors whose resistance is strongly dependent on temperature.

Some ear tags for feedlot cattle, which main technology is an accelerometer, also include a temperature sensor to gather additional data on animal status. However, rumen boluses to monitor ruminal temperature are probably the main PLF application of this type of sensors. Changes in ruminal temperature enable the identification of fever and drinking bouts. These sensors have also been used in beef cattle to detect calving and oestrus (Cooper-Prado *et al.*, 2011), bovine respiratory disease (Timsit *et al.*, 2011) and ruminal acidosis (Wahrmund *et al.*, 2012).

The use of temperature ruminal boluses is a challenge in the case of sheep because of the size of the bolus (a minimum volume is needed to include at least the microprocessor, the battery, and the communication system). Nonetheless, a small-sized rumen bolus (22×90 mm), suitable to be orally administered to small ruminants, has recently been developed at the Autonomous University of Barcelona (Caja *et al.*, 2020).

3.3.8 Weighing systems

Live weight is a key indicator to monitor the production efficiency of beef cattle and sheep, both in breeding and fattening production phases. There are several PLF solutions available for the monitoring of animal weight, which differs in the way weight data are gathered and the utilisation of those data, as described below.

Autodrafters

An autodrafter is a quite simple PLF solution, which can be described as the combination of a weighing crate (i.e. a scale) and a selective gate with the ability to distinguish and direct the passage of animals. Most times autodrafters have an RFID reader for the identification of each animal. They are quite common in large sheep farms, although specific versions for beef cattle are also available. They are mostly used to measure the weight of every single animal in a herd, redirecting lower weight animals towards specific pens, where supplementation is provided. Autodrafters are also useful to reduce the labour requirements of some farm activities, such as sheering, parasite treatments, or selection for sale. Morgan-Davies *et al.* (2018) observed that the proportion of lambs needing anthelmintic treatment was significantly reduced by 40% when autodrafters were used, resulting in a decrease of 46% in the amount of anthelmintic used. The total amount of labour required was also reduced by 36%.

Smart scales

A smart scale is simply a weighing crate that has been customised to automatically measure the live weight of fattening cattle. Typically, these crates are placed at watering troughs, using an RFID antenna for the identification of each animal within the pen. Every time an animal drinks, live weight is recorded by the system, enabling the continuous monitoring of growth rate, which is a key indicator for the improvement of decision-making in feedlots. When the smart scale is placed at the watering point, drinking behaviour can also be monitored, which may be used as a health indicator (Buhman *et al.*, 2000).

Several smart scales are commercially available. Some of them use short weighing platforms to register the weight of front legs, from which live weight can be estimated. These short-platform systems are often portable, enabling their use in several pens of the same feedlot. Commercial solutions are mostly focused on feedlot cattle, although smart scales specifically designed to be used under grazing conditions are also available. In this case, solar panels are used for energy supply and long-range communication protocols for data transmission.

Walk over weighing systems

These weighing systems include a one-way passage leading to a key stimulant (supplementary feeder, water, etc.) which the animals are forced to pass through, a weighing platform placed in that corridor, and an RFID antenna for the electronic identification of each animal. Continuous weight data are gathered by the system and body mass is estimated using averaging techniques (Long et al., 1991). Some authors concluded that walk over weighing systems (WoW) lacks repeatability for decision making (Brown et al., 2014), whereas others advocate for its use to provide regular information on animal weight change across an entire flock (Morris et al. 2012). Dickinson et al. (2013) obtained high agreements between WoW and static weight measurements in a short assay with dairy cows but reported low repeatability of WoW when compared to static weighing. This repeatability is even lower in the case of beef cattle and sheep, as these animals tend to pass through the corridor faster than dairy cows. Thus, González-García et al. (2018) designed a customised metallic structure with an S shape to be placed at the entrance of the WoW platform. This structure was designed to control the flow of animals and to force them (ewes in their study) to cross the platform with sufficient succession distance to prevent congestion.

Walk over weighing systems are extensively used by beef cattle and sheep farmers in North America and Oceania, but their use is limited in Europe, probably due to the smaller farm and herd size in EU countries.

Although weight control is the main application of WoW systems, they have also been used to monitor the pregnancy status of suckler cows and to determine parturition date based on the changes in weight profiles (Menzies et al., 2018).

3.3.9 Feed intake monitoring systems

Feed intake data are among the most desirable for farmers, particularly for meat producers. Together with live weight and growth rate, feed intake can be used to compute feed efficiency, which is a major indicator of the profitability and sustainability of livestock farming systems. However, individual feed intake data are hard to gather under farm conditions, as individual feed troughs are not common. From a PLF perspective, three main tools are available: feed bunk scanners, feed bunk readers and smart feeders.

Feed bunk scanners

Feed intake monitoring can be done automatically by scanning the amount of feed in a trough. Bunk scanning systems use a lidar scanner placed on top of the feeding trough to continuously monitor the quantity of remaining feed. Feed disappearence over time is a measure for the feed intake of the group.

Feed bunk readers

Feeding time and feeding behaviour (i.e. the total time each animal is located at the feed bunk, and how this time is distributed along a day) are related to feed intake (De Mol *et al.*, 2016). Based on that, commercial PLF solutions using RFID technology to register the presence or absence of each animal at the feed bunk are available. Most times these solutions have alarm systems for the identification of animals with a low presence at the feed bunk, as an indicator of health or welfare problems (Sowell *et al.*, 1998). However, the use of feeding time data to compute feed intake of single animals is not usual, as many other factors can affect this trait, e.g. those related to feed characteristics.

Smart feeders

The best option to accurately measure the feed intake of single animals is the use of smart feeders. They consist of individual feed bins placed on load bars to continuously monitor feed weight disappearance, which includes an electronic identification system, normally based on RFID, aimed at identifying each animal accessing the bin and recording feeding behaviour. These feeders have been commercially available for a long time, but their use is almost exclusive to research facilities and genetic centres due to cost limitations. Nowadays, despite the interests of fatteners in feed intake data, there is no low-cost option available to be used on commercial farms.

There are also smart feeders specifically designed to provide supplementary feeding to beef cattle and sheep on pasture. They have a specially designed food dispensing mechanism so that the use by specific animals can be controlled, i.e. specified animals in a herd could be allowed to use the smart feeder, while other animals could be excluded from using the system. Besides, an animal's intake can be controlled daily, so that after a set amount of feed is received for the day, the animal is no longer allowed to receive more feed. At breeding and research centres, smart feeders are sometimes coupled with greenhouse gas emissions monitoring devices.

3.3.10 Image-based monitoring systems

Theoretically, PLF systems based on image analysis could substitute the 'farmer's eye', as they enable continuous observation of animal body characteristics (size, conformation, etc.) and animal behaviour. However, there are some limitations for the use of cameras under commercial farm conditions, such as rats damaging the electric cables, flies and dirt obscuring camera lenses, etc. (Banhazi et al., 2015). Additionally, image analysis algorithms are often complex and need extensive research/calibration effort, as well as high computational capacities. In the case of rangelands, the use of camera systems is limited to their situation on strategic points, such as water points (Finch et al., 2006), or their fitting to drones and other aerial platforms (Mufford et al., 2019).

There are several types of camera systems, providing different features for the development of PLF applications, which are described in the following sub-sections.

RGB cameras

Two-dimensional images have been used for various applications related to precision beef and sheep farming, such as calving detection based on monitoring the locomotion and posture of pregnant cows (Cangar et al., 2008) or automatic classification of sheep breeds (Jwade et al., 2019). Nonetheless, the most common application of computer vision, which is commercially available for some beef cattle breeds, is gathering body measurements and estimating body weight (Ozkaya et al., 2016).

A newer but promising application of computer vision is the welfare assessment of single animals. For example, facial expressions of sheep have been analysed on images for the estimation of pain (Mahmoud et al., 2018).

3D cameras

Stereo cameras, also known as 3D cameras, are a type of camera with two or more lenses with a separate image sensor or film frame for each lens. This allows the camera to simulate human binocular vision and therefore gives it the ability to capture three-dimensional images.

As standard cameras, 3D cameras can be calibrated to measure body part dimensions or to study animal postures. Their most common use is the estimation of live weight and body condition score (BCS), although they have also been applied to the measurement of lameness score and other behaviour-related traits. They are relatively common in dairy commercial operations for the evaluation of BCS, but there are no commercial products for beef cattle and sheep farms.

Concerning beef cattle, several research studies have shown the potential of 3D cameras to estimate live weight, as well as to predict carcass characteristics of live finishing beef cattle (Miller *et al.*, 2019). While video image analysis is increasingly being used to grade carcasses in the abattoir, most producers still subjectively select animals for slaughter by the visual assessment of fat and condition score and/or by manual weighing. 3D imaging technology has the potential to provide predictions of carcass characteristics from live animals on the farm, allowing farmers to send cattle to slaughter as soon as they are within the parameters specified by the abattoir.

In the case of sheep, a weighing system based on 3D imagery (OtoP-3D) is under development as a collaboration project between the Institute National of Research for Agriculture, Nutrition and Environment (INRAe) and the Livestock Institute (IDELE) of France. The objective of this project is the automation of the evaluation of phenotypes to improve both production efficiency and genetic selection. Further research is needed to validate the use of this technology under farm conditions. Wool coverage may be a major limitation for its application in some sheep breeds, as it highly affects animal volumetry.

Thermal cameras

A body with a temperature above absolute zero emits radiation, forming an electromagnetic spectrum that can be absorbed by other bodies around it. It is possible to detect this kind of radiation with the use of thermal cameras, identifying even minimal temperature variations. Infrared thermography is a non-invasive sensing method used in measuring changes in heat transfer and blood flow, through the detection of small changes in body temperature.

Potential uses of infrared thermography in animal production are very diverse, as reviewed by McManus *et al.* (2016). Thermographic images may indicate changes in the blood flow resulting from an increased body temperature related to stressful environmental conditions. Paim *et al.* (2012) observed that the temperatures of the muzzle, neck, and rump were good indicators of the environmental and thermal comfort conditions of lambs. Infrared thermography has also been used in animal nutrition, e.g. as a tool to predict methane production and emission. Studies on energy metabolism have shown that beef production is more efficient with lower heat loss and methane production. The distal portions of the hindquarters and face temperatures are the most appropriate body sites to indirectly assess feed efficiency in cattle using thermography (Montanholi et al., 2009). Thermal images have also been used to estimate bull live weight, wither height and hip height (Stajnko et al., 2008).

Regarding animal health, temperature changes in a body region are in most cases the result of infection or inflammation, which are linked to vasodilation, making thermal imaging a diagnostic tool. Thermal imaging has been used in many veterinary applications, e.g. the detection of infective diseases, such as hoof lesions (Byrne et al., 2019), bluetongue (Pérez de Diego et al., 2013) or mastitis (Castro-Costa et al., 2014). Another interesting use of thermography, which is specific for extensive farming systems, is the inclusion of thermal cameras on drones to locate lost lambs during the lambing season (Caja et al., 2020).

Despite the potential of thermal cameras as PLF tools, there are some limitations and factors that must be considered when using infrared thermography. Images that are taken in sunlight or high humidity conditions, with convective heat loss due to wind, or when surfaces are dirty, must be carefully analysed. The radiation measured by the camera does not only depend upon the temperature of the object, as it is also a function of its emissivity and conductivity (Knízková et al., 2007). It is worth mentioning that those are very common situations under farm conditions, limiting the practical application of thermal cameras for precision farming applications.

3.3.11 Pasture monitoring systems

Technological solutions for monitoring grass availability and growth rate (quantity), as well as for monitoring grass nutritional quality are included in this sub-section.

Pasture quantity monitoring

Conventional methods to monitor grass production and availability include field measurements and statistics. Field measurements often refer to biomass harvesting, but eddy covariance tower measurements, field spectrometers or phenocams (digital cameras capturing time-lapse images of foliage), among others, are also included.

Probably, the best example in Europe of the utility of a grass availability monitoring service is PastureBase. It is an internet-based management programme for grassland farmers in Ireland, which is in operation since 2013. The service is based on data provided by farmers, who walk their paddocks weekly to measure the amount of grass. They can use several methods to estimate the quantity of grass in each paddock. The first option is to use a 0.5 × 0.5 m quadrat, which is placed in an area that is representative of the amount of grass in the paddock. The grass within the quadrat is cut to between 3.5 and 4 cm, which is considered the ideal post-grazing residual height, and dry matter yield is calculated by using a simple regression equation. Another option is to use a plate meter, which measures grass height. In this case, 4 cm are subtracted from the average height of the grass in the paddock and the difference is multiplied by 250 to get dry matter per hectare.

A first attempt to automate field monitoring is the use of electronic plate meters to measure sward height. These devices are raising plate meters with an ultra-sonic mounted sensor that recognises the distance of plate lift and can georeference each measurement with an integrated GNSS-receiver module. Electronic plate meters enable a faster and more precise measurement of grass height, as well as the automatic transmission of data to a cloud server, where they are standardised and analysed to support decision-making. Technological solutions equivalent to electronic plate meters, but based on grass capacitance, are also available in the market. In both cases, these tools require calibration models to estimate grass quantity from sensor data. Current models have demonstrated poor performance outside Ireland (Hart et al., 2019).

In addition to those field tools, green biomass can be monitored continuously using its spectral reflectance properties acquired by remote optical sensors, which could be mounted on drones or satellites. Optical sensors can be used to monitor the greenness, vitality, and density of vegetated areas. The utilisation of satellite data is of special interest when large and/or remote areas need to be monitored, as is the case of beef cattle and sheep farming. When using satellite information, the spatial, temporal, and spectral resolution must be considered. Spatial resolution depends primarily on the instantaneous field of view, which is the angular cone of visibility of the sensor and determines the area on the Earth's surface which is 'seen' from a given altitude at one particular moment in time. This area on the ground is called the resolution cell or pixel and determines a sensor's maximum spatial resolution. Temporal resolution refers to the temporal frequency of sampling by repeat imaging. Spectral resolution describes the ability of the sensor to define fine wavelength intervals. The finer the spectral resolution, the narrower the wavelength range for a particular channel or band.

Concerning spaceborne monitoring systems, apart from multi-spectral optical sensors, such as Advanced Very High-Resolution Radiometer (AVHRR), MODIS, Landsat, and Sentinel-2, there are also hyperspectral sensors available, such as Hyperion or CHRIS/Proba, which have the advantage of a higher spectral resolution. Some of these remote sensors provide open data, which can be used to gain information on the quantity of grass on large spatial scales, partly in an automated way. One of the main advantages of satellite data is the possibility to monitor intra-field variability, which is a key indicator of the adequacy of grazing management.

Vegetation indices, which are calculated from the values of optical sensor bands, are the most common proxy to the spatial and temporal patterns of grassland production (Reinermann et al., 2020). Several indices rely on the near-infrared and red bands, among others. The Normalised Difference Vegetation Index (NDVI) is by far the most used index, regarding pasture biomass monitoring (Punalekara et al., 2018).

A limitation for the use of satellite imagery to monitor grassland production and condition is the existence of data gaps due to clouds, which are especially common in northern Europe. The use of Synthetic Aperture Radar (SAR) sensors, such as Sentinel-1, to derive information on vegetation height and canopy structure can help to solve that problem, as radar signals are not affected by clouds. Further research is needed in this field.

Pasture quality monitoring

Concerning the monitoring of grass quality, Near-Infrared Reflectance Spectroscopy (NIRS) is the most promising technology available. NIRS is a vibrational spectroscopic technique based on the interaction of light with the material to obtain information on its composition, structure, microbiological contamination, or even parameters related to its organoleptic properties. NIRS enables rapid, accurate, non-destructive, and environmentally friendly analysis, together with the measurement of multiple parameters in a high variety of products. Thus, near-infrared spectral sensors combined with data analytics offer the possibility to provide cost-effective and added value solutions to a range of agriculture, livestock, and food problems.

The potential of NIRS for applications related to livestock production is enormous. In recent times, the instrumental development of NIR sensors has undergone an important revolution aimed at the development of smaller, portable sensors with a more stable signal, suitable for being used *in situ*. This enables measurements to be taken directly in the field, and therefore allows real-time decision-making. Besides, the combination of the spectral signal with other information and communication technologies opens enormous expectations in the application of NIR sensors to build decision-support systems.

Particularly, the use of NIRS is very widespread in the field of animal nutrition. There are some studies focused on the use of mobile NIRS analysis for monitoring pasture nutrients in real-time (Bell *et al.*, 2018), enabling a reduction in the time needed for analyses (from about one day to a minute) and the associated costs. NIRS has also been applied to the *in situ* characterisation of nutritional and ingredient composition of compound feeds, forages, or total mixed rations (Garrido *et al.*, 2016; Pérez *et al.*, 2004). It is worth mentioning other relevant applications of NIRS in livestock farming, such as the evaluation of the quality of animal products, e.g. meat, wool or milk (De la Roza *et al.*, 2017), or its use in carcasses or directly in live animals for the evaluation of fatness (for example in Iberian pigs), for authenticating the breed, or for predicting animal response (Pérez *et al.*, 2004; Pérez-Marín *et al.*, 2009).

Nonetheless, extracting relevant information from complex, and often high-dimensional sensor signals requires expertise to develop quantitative and qualitative prediction models that can be applied in the future to unknown samples (Pérez-Marín and Garrido, 2021). Robust and accurate models are essential to make this technology successful. Big data algorithms for multivariate data treatment must be applied to obtain robust predictive models.

A summary of the PLF systems for beef cattle and sheep described in this section can be found in Table 3.1 and 3.2, including the websites of commercially available systems.

Table 3.1. Summary of wearable precision lifestock farming technologies for beef cattle and sheep.

PLF technology	Device	Recorded traits	Applications	Species	Commercial product	Company websites
Radio Frequency Identification (RFID)	injectable, ear tag, and ruminal bolus	presence	individual identification	beef cattle and sheep	yes	allflex.global shearwell.co.uk datamars.com
Global Positioning Systems (GPS)	ear tag and collar	geographical position	location, theft control, behaviour monitoring, geofencing and virtual fencing	beef cattle and sheep	yes	brincow.com digitanimal.com domodis.com herdsy.com noldus.com nofence.no eshepherd.com.au
Accelerometers	ear tag, collar, and ruminal bolus	motion and position	behaviour monitoring, oestrus and calving detection, and lameness and acidosis monitorisation	beef cattle and sheep	yes	sensehubbeef.com quantifiedag.com
Pedometers	leg bracelet	step count	oestrus detection and lameness monitoring	beef cattle and sheep	yes	absglobal.com
Proximity loggers	harness	interactions between animals	oestrus detection and mounting activity monitoring	sheep	no	-
Microphones	halter	sound	bite and chew count	beef cattle and sheep	no	
Thermistors	ruminal bolus	temperature	eating and drinking behaviour	beef cattle and sheep	yes	moonsyst.com

Table 3.2. Summary of non-wearable precision lifestock farming technologies for beef cattle and sheep.[1]

PLF technology	Placement	Recorded traits	Applications	Species	Commercial product	Company websites
Autodrafters	alley	weight	automatic sorting based on live weight	sheep	yes	shearwell.co.uk ritchie-d.co.uk
Smart scales	water trough	weight and drinking behaviour	live weight and growth monitoring, and detection of health problems	beef cattle	yes	digitanimal.com ritchie-d.co.uk hen-col.com c-lockinc.com growsafe.com optiweigh.com.au
Walk over weighing systems (WoW)	alley	weight	live weight monitoring	beef cattle and sheep	yes	livestock.tru-test.com
Feed bunk scanners	feed bunk	amount of feed	group feed monitoring	beef cattle	yes	septentrio.com
Feed bunk readers	feed bunk	presence	health and wellbeing monitoring	beef cattle	yes	anitrace.net
Smart feeders	feed bunk	feed intake	feed efficiency monitoring	beef cattle	yes	c-lockinc.com growsafe.com
RGB cameras	barn camera and smartphone	RGB image	weight estimation and welfare monitoring	beef cattle and sheep	Yes	agroninja.com
3D cameras	barn and alley	volumetry	BCS and weight control, and carcass yield estimation	beef cattle and sheep	No	-
Thermal cameras	barn and portable device	surface temperature	health monitoring and feed efficiency estimation	beef cattle and sheep	no	-
Microsonics	portable device	sward height	herbage mass monitoring	beef cattle and sheep	yes	moregrass.ie
Capacitive sensors	portable device	capacitance	herbage mass monitoring	beef cattle and sheep	yes	novelways.nz
Spectral sensors	UAV or satellite-borne	reflectance	herbage mass monitoring	beef cattle and sheep	no	-
Near-Infrared Reflectance Spectroscopy (NIRS)	portable device	reflectance	grass quality monitoring	beef cattle and sheep	yes	abvista.com grainit.it

[1] UAV = unmanned aerial vehicle.

3.4 Future trends in precision beef and sheep farming

3.4.1 Enhanced traceability

Currently, most PLF applications are aimed at improving efficiency through the reduction of costs and the optimisation of productivity. Nevertheless, PLF applications can also be tools for gathering unbiased data about farming practices. This type of data could be accessible by consumers as a certification of some best practices or high production standards, i.e. PLF may provide enhanced traceability systems. A good example of this potential application is welfare monitoring. While many PLF applications are focused on the early detection of welfare issues, consumers are increasingly interested in animal welfare certification schemes. Several projects are currently working on the adaptation of PLF tools to the evaluation of animal welfare under international standards, such as Welfare Quality®, and it is feasible to have more of these applications in the future (Maroto-Molina *et al.*, 2020).

Regarding traceability, PLF could have a huge impact on the future of extensive beef cattle and sheep production systems. It is important to highlight that grasslands cover one-third of the Earth's terrestrial surface and that, in some European countries, e.g. Ireland, they account for 90% of the agricultural land area. Apart from providing forage for livestock production, grasslands fulfil several functions and ecosystem services, which make them essential. The most important ones are carbon storage, biodiversity, water purification, erosion control, and recreation. The EU Green Deal is a set of policy initiatives by the European Commission with the overarching aim of making Europe climate neutral in 2050. Livestock grazing is considered one of the tools to reach that objective and has been proposed as an eco-scheme in several EU countries. As a future CAP innovation, eco-schemes shall provide support for farmers who observe agricultural practices beneficial for the environment and climate. It aims to be a measure to reward and incentivise farmers for acting towards a more sustainable farm and land management to maintain public goods. However, to be eligible for payment, farmers will need to demonstrate that their animals graze sustainably, avoiding both under- and overgrazing. Some PLF technologies, such as GPS collars, probably coupled with satellite imagery, have an enormous potential to gather information on grazing patterns. Thus, it is expected that PLF tools for the certification of ecosystem services associated with beef cattle and sheep farming will be developed.

3.4.2 Energy supply

Several PLF applications for beef cattle and sheep include sensors and communication protocols that demand a high energy supply, e.g. GPS collars, especially when they are used for virtual fencing. This leads to the need for batteries that are either large or with short life spans, limiting the application of these technologies under commercial conditions. Some PLF tools have incorporated solar panels on collars, but it makes devices more susceptible to bumps or scratches and it only provides a slightly longer life span (current commercial solutions promise three months without changing batteries). New developments on energy storage efficiency (i.e. the capacity to store more energy in smaller and lighter batteries), as well as on reduced energy needs of sensors and communication services, are expected in the upcoming years. Another promising field of research is the harvest of energy (chemical, thermal, or mechanical) from the animals' body, allowing the existence of self-powered electronics (Dagdeviren *et al.*, 2017).

3.4.3 Robotics

The existence of physically exerting or boring tasks associated with beef cattle and sheep farming, such as shearing or herding, is a limitation for the incorporation of young farmers to these sectors, especially in the case of extensive farming, where some tasks must be carried out under uncomfortable climate conditions. Robots offer the possibility to automate some farmers tasks, improving their quality of life and the attractiveness of farming as a profession. One example of possible future developments is robotic herding, i.e. the use of robots to move sheep flocks to a certain location. At the moment, few trials have been made using herding robots, e.g. SPOT, the four-legged robot of Boston Dynamics which was developed mainly for military and industrial purposes. Although some studies showed that animals get habituated to robots, it does not necessarily mean that the intelligent robotic movement of sheep flocks is impractical. An interesting possibility being investigated is that sheep may even accept a robot into the flock in such a way that, with some form of positive reinforcement, they can learn to view it as a leader and simply follow it to a new location (Evered *et al.*, 2014).

References

Alhamada, M., Debus, N., Lurette, A., Bocquier, F., 2017. Automatic oestrus detection system enables monitoring of sexual behaviour in sheep. Small Ruminant Research 149: 105-111.

Allen, V.G., Batello, C., Berretta, E.J., Hodgson, J., Kothmann, M., Li, X., McIvor, J., Milne, J., Morris, C., Peeters, A., Sanderson, M., 2011. An international terminology for grazing lands and grazing animals. Grass and Forage Science 66: 2-28.

Anderson, D.M., 2007. Virtual fencing – past, present, and future. The Rangeland Journal 29: 65-78.

Andriamandroso, A.L.H., Lebeau, F., Beckers, Y., Froidmont, E., Dufrasne, I., Heinesch, B., Dumortier, P., Blanchy, G., Blaise, Y., Bindelle, J., 2017. Development of an open-source algorithm based on inertial measurement units (IMU) of a smartphone to detect cattle grass intake and ruminating behaviours. Computers and Electronics in Agriculture 139: 126-137.

Banhazi, T., Vranken, E., Berckmans, D., Rooijakkers, L., Berckmans, D., 2015. Word of caution for technology providers: practical problems associated with large scale deployment of PLF technologies on commercial farms. In: Halachmi, I. (ed.) Precision livestock farming applications: making sense of sensors to support farm management. Wageningen Academic Publishers, Wageningen, the Netherlands, pp. 105-112.

Bell, M.T., Mereu, L., Davis, J., 2018. The use of mobile near-infrared spectroscopy for real-time pasture management. Frontiers in Sustainable Food Systems 2: 76. https://doi.org/10.3389/fsufs.2018.00076

Brown, D.J., Savage, D.B., Hinch, G.N., 2014. Repeatability and frequency of in-paddock sheep walk-over weights: implications for individual animal management. Animal Production Science 54: 207-213. https://doi.org/10.1071/AN12311

Buhman, M.J., Perino, L.J., Galyean, M.L., Wittum, T.E., Montgomery, T.H., Swingle, R.S., 2000. Association between changes in eating and drinking behaviours and respiratory tract disease in newly arrived calves at a feedlot. American Journal of Veterinary Research 61: 1163-1168.

Byrne, D.T., Berry, D.P., Esmonde, H., McGovern, F., Creighton, P.H., McHugh, N., 2019. Infrared thermography as a tool to detect hoof lesions in sheep. Translational Animal Science 3: 577-588.

Caja, G., Castro-Costa, A., Salama, A., Oliver, J., Baratta, M., Ferrer, C., Knight, C., 2020. Sensing solutions for improving the performance, health, and wellbeing of small ruminants. Journal of Dairy Research 87 (Suppl. 1): 34-46. https://doi.org/10.1017/S0022029920000667

Cangar, O., Leroy, T., Guarino, M., Vranken, E., Fallon, R., Lenehan, J., Mee, J., Berckmans, D., 2008. Automatic real-time monitoring of locomotion and posture behaviour of pregnant cows prior to calving using online image analysis. Computers and Electronics in Agriculture 64(1): 53-60.

Castro-Costa, A., Caja, G., Salama, A.A.K., Rovai, M., Flores, C., Aguiló, J., 2014. Thermographic variation of the udder of dairy ewes at early-lactation and following an *E. coli* endotoxin intramammary challenge at late lactation. Journal of Dairy Science 97: 1377-1387.

Cooper-Prado, M.J., Long, N.M., Wright, E.C., Goad, C.L., Wettemann, R.P., 2011. Relationship of ruminal temperature with parturition and oestrus of beef cows. Journal of Animal Science 89: 1020-1027. https://doi.org/10.2527/jas.2010-3434

Dagdeviren, C., Li, Z., Wang, Z.L., 2017. Energy harvesting from the animal/human body for self-powered electronics. Annual Review of Biomedical Engineering 19: 85-108.

De la Roza-Delgado, B., Garrido, A., Soldado, A., González, A., Cuevas, M., Maroto, F., Pérez-Marín, D., 2017. Matching portable NIRS instruments for *in situ* monitoring indicators of milk composition. Food Control 76: 74-81.

De la Torre, M.P., Briefer, E.F., Reader, T., McElligott, A.G., 2015. Acoustic analysis of cattle (*Bos taurus*) mother-offspring contact calls from a source-filter theory perspective. Applied Animal Behaviour Science 163: 58-68.

De Mol, R.M., Goselink, R.M.A., Van Riel, J.W., Knijn, H.M., Van Knegsel, A.T.M., 2016. The relation between eating time and feed intake of dairy cows. In: Kamphuis, C., Steeneveld, W. (eds.) Precision Dairy Farming 2016. Wageningen Academic Publishers, Wageningen, the Netherlands, pp. 387-392.

Dickinson, R.A., Morton, J.M., Beggs, D.S., Anderson, G.A., Pyman, M.F., Mansell, P.D., Blackwood, C.B., 2013. An automated walk-over weighing system as a tool for measuring liveweight change in lactating dairy cows. Journal of Dairy Science 96: 4477-4486.

European Statistics Office (EUROSTAT), 2021. Available at: https://ec.europa.eu/eurostat/web/main/data/database.

Evered, M., Burling, P., Trotter, M., 2014. An investigation of predator response in robotic herding of sheep. In: Proceedings of International Conference on Intelligent Agriculture. IACSIT Press, Singapore, pp. 49-54.

Federal Aviation Administration (FAA), 2021. Available at: https://tinyurl.com/b4p4uf7b.

Finch, N.A., Murray, P.J., Dunn, M.T., Billingsley, J., 2006. Using machine vision classification to control access of animals to water. Australian Journal of Experimental Agriculture 46: 837-839.

Fogarty, E.S., Swain, D.L., Cronin, G., Trotter, M., 2018. Autonomous on-animal sensors in sheep research: a systematic review. Computers and Electronics in Agriculture 150: 245-256.

Fogarty, E.S., Swain, D.L., Cronin, G.M., Moraes, L.E., Bailey, D.W., Trotter, M.G., 2020. Potential for autonomous detection of lambing using global navigation satellite system technology. Animal Production Science 60: 1217-1226.

Food and Agriculture Organisation (FAO), 2001. Pastoralism in the new millennium. FAO Animal Production and Health paper 150. Available at: https://www.fao.org/docrep/005/Y2647E/Y2647E00.HTM.

Galli, J.R., Cangiano, C.A., Pece, M.A., Larripa, M.J., Milone, D.H., Utsumi, S.A., Laca, E.A., 2018. Monitoring and assessment of ingestive chewing sounds for prediction of herbage intake rate in grazing cattle. Animal 12: 973-982.

Garrido, A., Vega, S., Maroto, F., de la Haba, M.J., Pérez-Marín, D., 2016. On-site quality control of processed land animal proteins using a portable micro-electro-mechanical-systems near infrared spectrometer. Journal of Near Infrared Spectroscopy 24: 47-58.

González-García, E., Alhamada, M., Pradel, J., Douls, S., Parisot, S., Bocquier, F., Menassol, J.B., Llach, I., González, L.A. 2018. A mobile and automated walk-over-weighing system for a close and remote monitoring of liveweight in sheep. Computers and Electronics in Agriculture 153: 226-238.

Hamilton, A.W., Davison, C., Tachtatzis, C., Andonovic, I., Michie, C., Ferguson, H.J., Somerville, L., Jonsson, N.N., 2019. Identification of the rumination in cattle using support vector machines with motion-sensitive bolus sensors. Sensors 19: 1165. https://doi.org/10.3390/s19051165

Hart, L., Oudshoorn, F., Latsch, R., Umstätter, C., 2019. How accurate is the Grasshopper system in measuring dry matter quantity of Swiss and Danish grassland? In: Proceedings of the 9th European Conference on Precision Livestock Farming (ECPLF). Moorepark, Fermoy, Co. Cork, Fermoy, pp. 188-193.

Hocquette, J., Ellies-Oury, M., Lherm, M., Pineau, C., Deblitz, C., Farmer, L., 2018. Current situation and prospects for beef production in Europe: a review. Asian-Australasian Journal of Animal Science 31: 1017-1035.

Ihle, R., Dries, L., Jongeneel, R., Venus, T., Wesseler, J., 2017. Research for agricommittee The EU cattle sector: challenges and opportunities – milk and meat. European Parliament, Directorate general for internal policies. Policy department B: structural and cohesion policies agricultural and rural development European Parliament Committees. Available at: https://tinyurl.com/ewk8f934.

Islam, M.A., Lomax, S., Doughty, A.K., Islam, M.R., Clark, C.E.F., 2020. Automated monitoring of panting for feedlot cattle: sensor system accuracy and individual variability. Animals 10: 1518. https://doi.org/10.3390/ani10091518

Jwade, S.A., Guzzomi, A., Mian, A., 2019. On farm automatic sheep breed classification using deep learning. Computers and Electronics in Agriculture 167: 105055.

Knízková, I., Kunc, P., Gürdíl, G.A.K., Pinar, Y., Selví, K.Ç., 2007. Applications of infrared thermography in animal production. Journal of the Faculty of Agriculture of Kyushu University 22: 329-336.

Lherm, M., Agabriel, J., Devun, J., 2017. Status and trends of suckler beef production in France and in three European countries. In: Agabriel, J., Renand, G., Baumont, R. (eds.) Suckler beef production. Vol. 30. Dossier INRA Productions Animales. Paris, France, pp. 93-106.

Long, J., Takahata, H., Umetsu, K., Hoshiba, H., Takeyama, I., 1991. A livestock walk-through scale system. Journal of the Society of Agricultural Structures 21: 175-182.

Mahmoud, M., Lu, Y., Hou, X., McLennan, K., Robinson, P. 2018. Estimation of pain in sheep using computer vision. In: Moore, R.J. (ed.) Handbook of pain and palliative care. Springer Nature, Basel, Switzerland.

Marini, D., Meuleman, M.D., Belson, S., Rodenburg, T.S., Llewellyn, R., Lee, C., 2018. Developing an ethically acceptable virtual fencing system for sheep. Animals 8: 33.

Maroto-Molina, F., Gómez-Cabrera, A., Guerrero-Ginel, J.E., Garrido-Varo, A., Adame-Siles, J.A., Pérez-Marín, D.C., 2018. Caracterización y tipificación de explotaciones de dehesa asociadas a cooperativas: un caso de estudio en España. Revista Mexicana de Ciencias Pecuarias 9: 811-832. https://doi.org/10.22319/rmcp.v9i4.4534

Maroto-Molina, F., Navarro-García, J., Príncipe-Aguirre, K., Gómez-Maqueda, I., Guerrero-Ginel, J.E., Garrido-Varo, A., Pérez-Marín, D.C., 2019. A low-cost IoT-based system to monitor the location of a whole herd. Sensors 19: 2298. https://doi.org/10.3390/s19102298

Maroto-Molina, F., Pérez-Marín, C., Molina-Moreno, L., Agüera-Buendía, E., Pérez-Marín, D., 2020. Welfare Quality® for dairy cows: towards a sensor-based assessment. Journal of Dairy Research 87 (Suppl. 1): 28-33. https://doi.org/10.1017/S002202992000045X

McManus, C., Tanure, C.B., Peripolli, V., Seixas, L., Fischer, V., Gabbi, A.M., Menegassi, S.R.O., Stumpf, M.T., Kolling, G.J., Dias, E., Costa, J.B.G., 2016. Infrared thermography in animal production: an overview. Computers and Electronics in Agriculture 123: 10-16.

Menzies, D., Patison, K.P., Corbet, N.J., Swain, D.L., 2018. Using walk-over-weighing technology for parturition date determination in beef cattle. Animal Production Science 58: 1743-1750.

Miller, G.A., Hyslop, J.J., Barclay, D., Edwards, A., Thomson, W., Duthie C.A., 2019. Using 3D imaging and machine learning to predict liveweight and carcass characteristics of live finishing beef cattle. Frontiers in Sustainable Food Systems 3: 30.

Montanholi, Y.R., Swanson, K.C., Schenkel, F.S., Mcbride, B.W., Caldwell, T.R., Miller, S.P., 2009. On the determination of residual feed intake and associations of infrared thermography with efficiency and ultrasound traits in beef bulls. Livestock Science 125: 22-30. http://dx.doi.org/10.1016/j.livsci.2009.02.022

Morgan-Davies, C., Lambe, N., Wishart, H., Waterhouse, T., Kenyon, F., McBean, D., McCracken, D., 2018. Impacts of using a precision livestock system targeted approach in mountain sheep flocks. Livestock Science 208: 67-76.

Morris, J.E., Cronin, G.M., Bush, R.D., 2012. Improving sheep production and welfare in extensive systems through precision sheep management. Animal Production Science 52: 665-670.

Morris, S.T., 2017. Overview of sheep production. In: Ferguson, D.M., Lee, C. and Fisher A. (eds.) Advances in Sheep Welfare. Woodhead Publishing, Cambridge, UK, pp. 19-35.

Mufford, J.T., Hill, D.J., Nancy, J.F., Church, J.S., 2019. Use of unmanned aerial vehicles (UAVs) and photogrammetric image analysis to quantify spatial proximity in beef cattle. Journal of Unmanned Vehicle Systems 7: 194-206. https://doi.org/10.1139/juvs-2018-0025

O'Leary, N.W., Byrne, D.T., O'Connor, A.H., Shalloo, L., 2020. Invited review: cattle lameness detection with accelerometers. Journal of Dairy Science 103: 3895-3911.

O'Neill, C.J., Bishop-Hurley, G.J., Williams, P.J., Reid, D.J., Swain, D.L., 2014. Using UHF proximity loggers to quantify male-female interactions: a scoping study of oestrous activity in cattle. Animal Reproduction Science 151: 1-8.

Ozkaya, S., Neja, W., Krezel-Czopek, S., Oler, A., 2016. Estimation of body weight from body measurements and determination of body measurements on Limousin cattle using digital image analysis. Animal Production Science 56: 2060-2063.

Paim, T.P., Borges, B.O., Lima, P.M.T., 2012. Relation between thermographic temperatures of lambs and thermal comfort indices. International Journal of Applied Animal Sciences 1: 108-115.

Patison, K., Trotter, M., Swain, D., Corbet, N., Bailey, D., Kinder, J., 2017., Applying proximity sensors to monitor beef cattle social behaviour as an indicator of animal welfare. In: Nelson W., McKenzie, L. (eds.) Proceedings of the 1st Asian-Australasian Conference on Precision Pastures and Livestock Farming. October 16-18, 2017. Hamilton, New Zealand, pp. 20.

Pérez de Diego, A.C., Sánchez-Cordón, P.J., Pedrera, M., Martínez-López, B., Gómez-Villamandos, J.C., Sánchez-Vizcaíno, J.M., 2013. The use of infrared thermography as a non-invasive method for fever detection in sheep infected with bluetongue virus. Veterinary Journal 198: 182-186.

Pérez, C., Rodríguez, I., Corral, S., Dorado, J., Hidalgo, M., Garrido, A., Pérez-Marín, D., 2004. Feasibility study for predicting progesterone levels in cattle plasma by NIRS. In: Davies T., Garrido A. (eds.) Proceedings of the 11th International Conference. NIR Publications. Chichester, UK, pp. 927-930.

Pérez-Marín, D., De Pedro, E., Guerrero, J.E., Garrido, A., 2009. A feasibility study on the use of near-infrared spectroscopy for prediction of the fatty acid profile in live Iberian pigs and carcasses. Meat Science 83: 627-633.

Pérez-Marín, D., Garrido A., 2021. NIR sensors for the in-situ assessment of Iberian ham. In: Cifuentes A. (ed.), Comprehensive Foodomics, vol. 3, Elsevier, Kidlington, Oxford, UK, pp. 340-345.

Powell, T.L., 1968. Pedometer measurements of the distance walked by grazing sheep in relation to weather. Grass and Forage Science 23: 98-102.

Punalekara, S.M., Verhoefa, A., Quaifeb, T.L., Humphriesc, D., Berminghamd, L., Reynoldsc, C.K., 2018. Application of Sentinel 2A data for pasture biomass monitoring using a physically based radiative transfer model. Remote Sensing of Environment 218: 207-220.

Reinermann, S., Asam, S., Kuenzer, C., 2020. Remote sensing of grassland production and management – a review. Remote Sensing 12: 1949.

Schleppe, J.B., Lachapelle, G., Booker, C., and Pittman, T., 2010. Challenges in the design of a GNSS ear tag for feedlot cattle. Computers and Electronics in Agriculture 70: 84-95.

Sheng, H., Zhang, S., Zuo, L., Duan, G., Zhang, H., Okinda, C., Shen, M., Chen, K., Lu, M., Norton, T., 2020. Construction of sheep forage intake estimation models based on sound analysis. Biosystems Engineering 192: 144-158.

Sowell, B.F., Bowman, J.G.P., Branine, M.E., Hubbert, M.E., 1998. Radio frequency technology to measure feeding behaviour and health of feedlot steers. Applied Animal Behaviour Science 59: 277-284.

Stajnko, D., Brus, M., Hočevar, M., 2008. Estimation of bull live weight through thermographically measured body dimensions. Computers and Electronics in Agriculture 61: 233-240.

Swain, D.L., Bishop-Hurley, G.J., 2007. Using contact logging devices to explore animal affiliations: Quantifying cow-calf interactions. Applied Animal Behaviour Science 102: 1-11.

Theurer, M.E., Amrine, D.E., White, B.J., 2013. Remote non-invasive assessment of pain and health status in cattle. Veterinary Clinics of North America: Food Animal Practice 29: 59-74.

Timsit, E., Assié, S., Quiniou, R., Seegers, H., Bareille, N., 2011. Early detection of bovine respiratory disease in young bulls using reticulo-rumen temperature boluses. The Veterinary Journal 190: 136-142.

Tomkiewicz, S.M., Fuller, M.R., Kie, J.G., Bates, K.K. 2010. Global positioning system and associated technologies in animal behaviour and ecological research. Philosophical Transactions of the Royal Society B 365, 2163-2176.

Triguero-Ocaña, R., Vicente, J., Acevedo, P. 2019. Performance of proximity loggers under controlled field conditions: an assessment from a wildlife ecological and epidemiological perspective. Animal Biotelemetry 7: 24.

Umstätter, C., 2011. The evolution of virtual fences: a review. Computers and Electronics in Agriculture 75: 10-22.

Valenza, A., Giordano, J.O., Lopes, G., Vincenti, L., Amundson, M.C., Fricke, P.M., 2012. Assessment of an accelerometer system for detection of oestrus and treatment with gonadotropin-releasing hormone at the time of insemination in lactating dairy cows. Journal of Dairy Science 95: 7115-7127.

Wahrmund, J.L., Ronchesel, J.R., Krehbiel, C.R., Goad, C.L., Trost, S.M., Richards, C.J., 2012. Ruminal acidosis challenge impact on ruminal temperature in feedlot cattle. Journal of Animal Science 90: 2794-2801. https://doi.org/10.2527/jas.2011-4407

Wolfger, B., Timsit, E., Pajor, E., Cook, N., Barkema, H., Orsel, K., 2015. Technical note: accuracy of an ear tag-attached accelerometer to monitor rumination and feeding behaviour in feedlot cattle. Journal of Animal Science 93: 3164-3168.

4. Precision pig farming

P. Jacobs and E. van Erp-van der Kooij

HAS University of Applied Sciences, P.O. Box 90108, 5200 MA 's Hertogenbosch, the Netherlands; l.verp@has.nl

Highlights

- The pig sector consists of breeding and finishing or closed farms, where sows produce piglets that grow in approximately 6 months to slaughter weight.

- Sows produce 2.5 litters per year with an average of 14 piglets per litter.

- The general trend is that the number of farms decreases but the farm size increases, and less labour is needed per pig.

- The development of smart farming techniques to provide extra eyes and ears in the pig units is increasing, to help to ensure sufficiently high animal welfare, health, and production performance levels.

- Integral data collection, clever data storage and dashboards that visualise data provide information to the farmer to improve results.

- Vision and sound are promising technologies in the pig sector.

- Devices must be robust and systems should be integrated to be implemented on a larger scale.

4.1 The pig production chain

Pig farming is focused on meat production. At the top of the breeding pyramid, on nucleus farms, sows and boars are selected for reproduction (sow lines) or growth (boar lines). From these purebred lines, crosses are made, that deliver the sows to the farms and the boars to the artificial insemination (AI) stations to produce the semen for the breeding sow farms. In specialised breeding farms, sows produce piglets that are grown to slaughter weight on finishing farms. The structure of the sector is shown in Figure 4.1.

Breeding companies work with nucleus farms to breed the best pigs for the next generation. These companies use data and information of the individual animals and their results to calculate the breeding value of the pigs and select the best parents to obtain the maximum genetic progress in the offspring. PIC is the largest pig breeding company worldwide, followed by Topigs Norsvin.

4.1.1 The pig farm

Three types of farms can be distinguished in specialised pig farming: pig breeding companies, finishing pig farms, and closed (or combined) companies. Pig breeding companies house sows that produce piglets. Sows produce about 2.5 litters per year with a mean number of 14 live-born piglets per litter (Agrisyst), resulting in

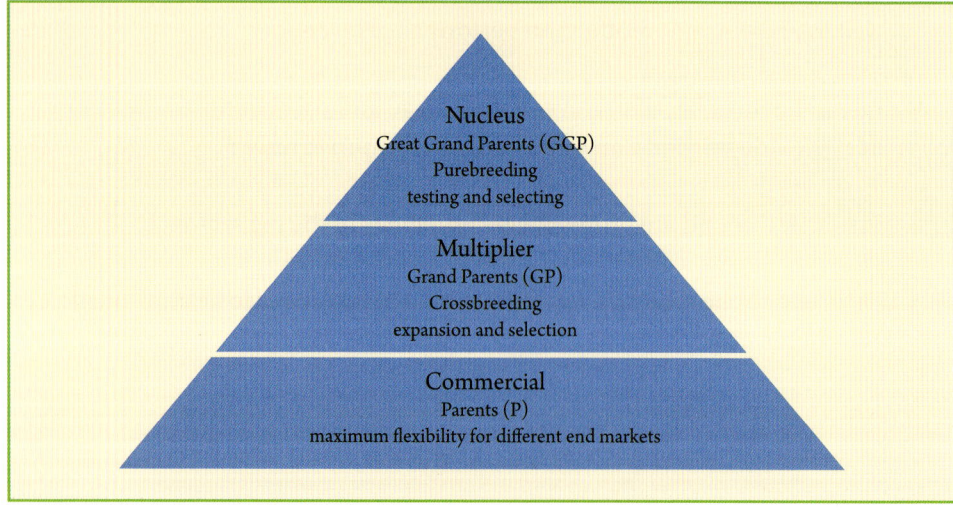

Figure 4.1. The pig breeding pyramid (modified from Meerburg and Pest, 2014).

approximately 35 live born and 29.4 piglets weaned per sow per year in 2019; in 2002 this was 22.1 weaned piglets per sow per year (Van der Meulen, 2020a). Finishing farms house pigs from 25 kg (appr. 10 weeks of age) to slaughter weight (appr. 5.5 months of age). Closed pig farms produce pigs from birth until slaughter, so they have sows, piglets, weaned pigs, and finishing pigs. In closed farms, no piglets are brought into the farm, only replacement sows and sperm to fertilise the sows.

4.1.2 Production structure in historical perspective

In April 2020, the Netherlands had a total of 11.9 million pigs (Figure 4.2) on approximately 3,500 farms (Figure 4.3), operated by approximately 3,500 entrepreneurs. The number of pig farms in the Netherlands is gradually decreasing. However, the total number of pigs remains approximately the same and therefore the farms are becoming on average larger. In Europe, other important pig (meat) producing countries are Germany, Spain, France, Denmark, and Poland.

The 11.9 million pigs in 2020 consist of more than 870,000 sows, 5.4 million piglets (either with the sow or weaned piglets), and 5.4 million finishing pigs (Van der

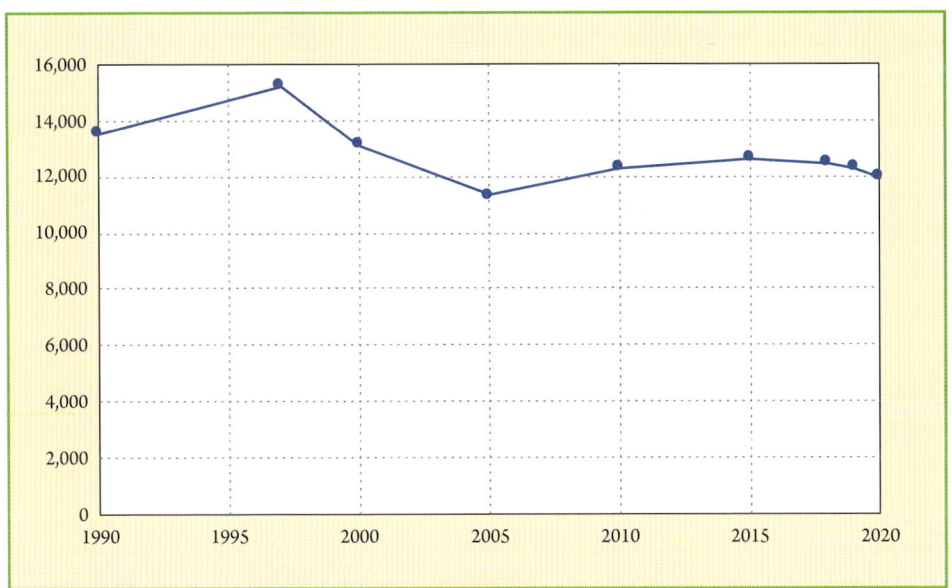

Figure 4.2. Historical development of the number of pigs in the Netherlands (×1000) (redrawn from CBS, 2020).

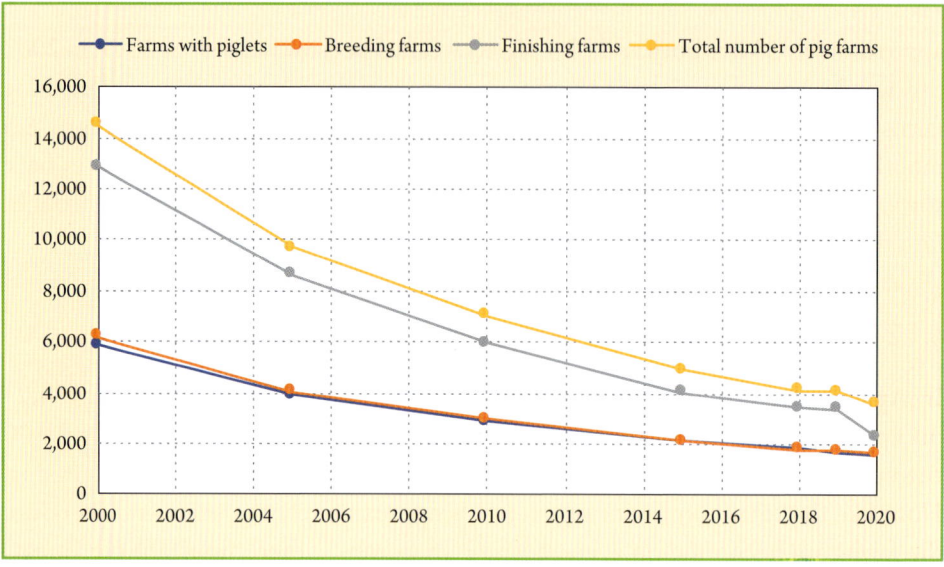

Figure 4.3. Historical development of the number of farms with pigs in the Netherlands (redrawn from CBS, 2020).

Meulen, 2020b). In the last 20 years, the number of pig farms has decreased from appr. 14,500 to 3,500 farms: a decrease of >75% (European Commission, 2020).

Pig farming has traditionally been part of the mixed farm system, where farms combined arable and animal production. The surface, parcelling or soil conditions of the cultivated land have often been limiting factors for expansion in cattle farming or arable farming. The advantages of the mixed farm included spreading the business risks over multiple production directions, a flexible business structure, and adding value to (by) products from dairy farming and arable farming.

The strong production expansion of pig farming in the years before 1980 in the Netherlands largely took place on the mixed farms in the sandy soil areas. For these farms, which otherwise might have had to be closed because they were no longer of sufficient size, intensive livestock farming, with the predominantly overseas supplied raw feed materials, offered an opportunity to increase production volume and income.

The production structure of the mixed farm changed over the years in the direction of specialisation per animal species and a simultaneous increase in scale. Because no land is needed for the feed supply of the pigs (the feed is usually produced in a mill), pig farming has developed into an industry that is mostly not land related.

The advantage of a specialised and large-scale company lies in the more efficient use of the production resources, which means that cost savings can occur, and yields can be increased. Mandatory hygiene and welfare investments are more difficult to afford by smaller companies. Specialisation allows the entrepreneur's professional competence to develop further.

4.1.3 Regional location of the companies

Pig farming in the Netherlands, like the other intensive livestock farming sectors, is strongly concentrated on farms in the sandy areas in the south and east (Figure 4.4). Noord-Brabant and Gelderland are the most important pig provinces. Within the sandy soil areas, the centre of gravity has gradually shifted from Gelderland to Noord-Brabant. The shares of Overijssel and Limburg have remained fairly constant. The highest number of pigs per hectare can be found in the Southeast Brabant region, with the municipality of Venray as the most pig-dense area: almost 660,000 pigs live there on 111 farms: approximately 71 pigs per hectare. The percentage of farms approaches the percentage of pigs per province. Noord-Brabant has about 44% of the Dutch pigs on 40% of the farms, while Gelderland has 20% of the pigs on 25%

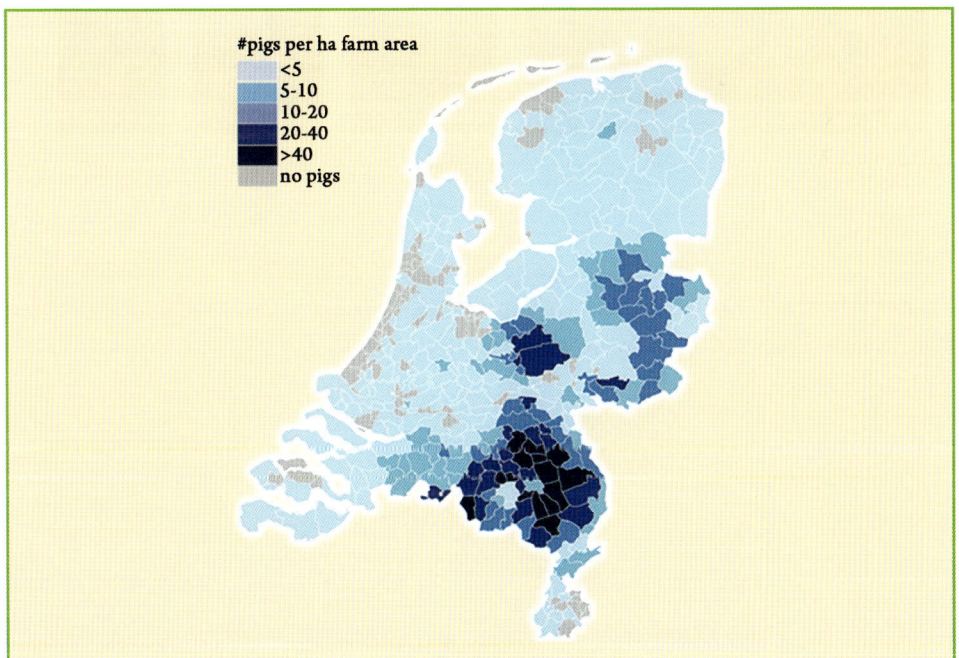

Figure 4.4. Pig density per municipality (CBS, http://www.cbs.nl).

of the farms. Besides regular or conventional farms, there is a niche of organic farms that houses <1% of the total pig population on appr. 148 farms. Very large farms also exist: in 2013, there were 157 so-called 'megafarms' in the Netherlands, with >1,200 sows or >7,500 finishing pigs (Van der Peet *et al.*, 2018) and in 2016 there were 326 of such megafarms (CBS, 2020).

In Europe, the number of pigs per inhabitant can be seen in Figure 4.5. The distribution of pigs over Europe can be found in Figure 4.6.

The EU produced 22,8 million tons of pig meat in 2019. Germany and Spain were the largest producers in the EU, with respective shares of 23% and 20% of the production (Figure 4.7).

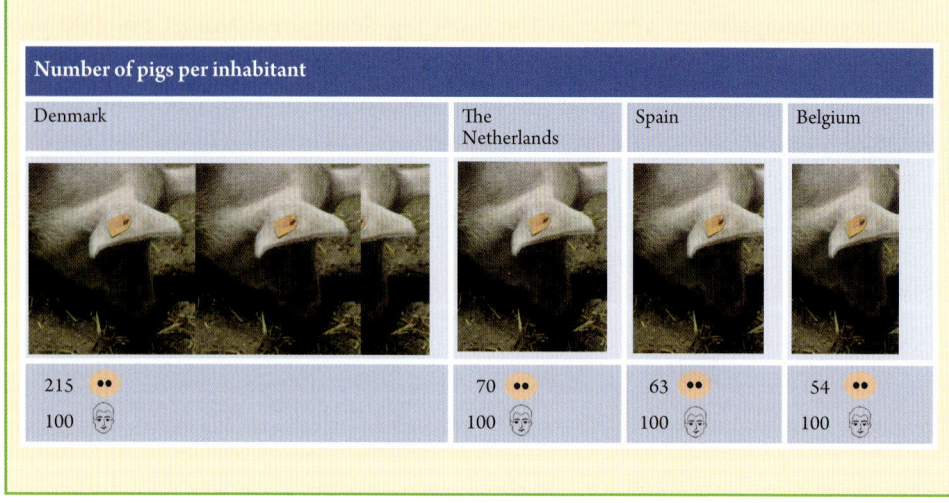

Figure 4.5. Number of pigs per inhabitant of Denmark, the Netherlands, Spain and Belgium (modified from European Union, 2020; http://www.ec.europa.eu).

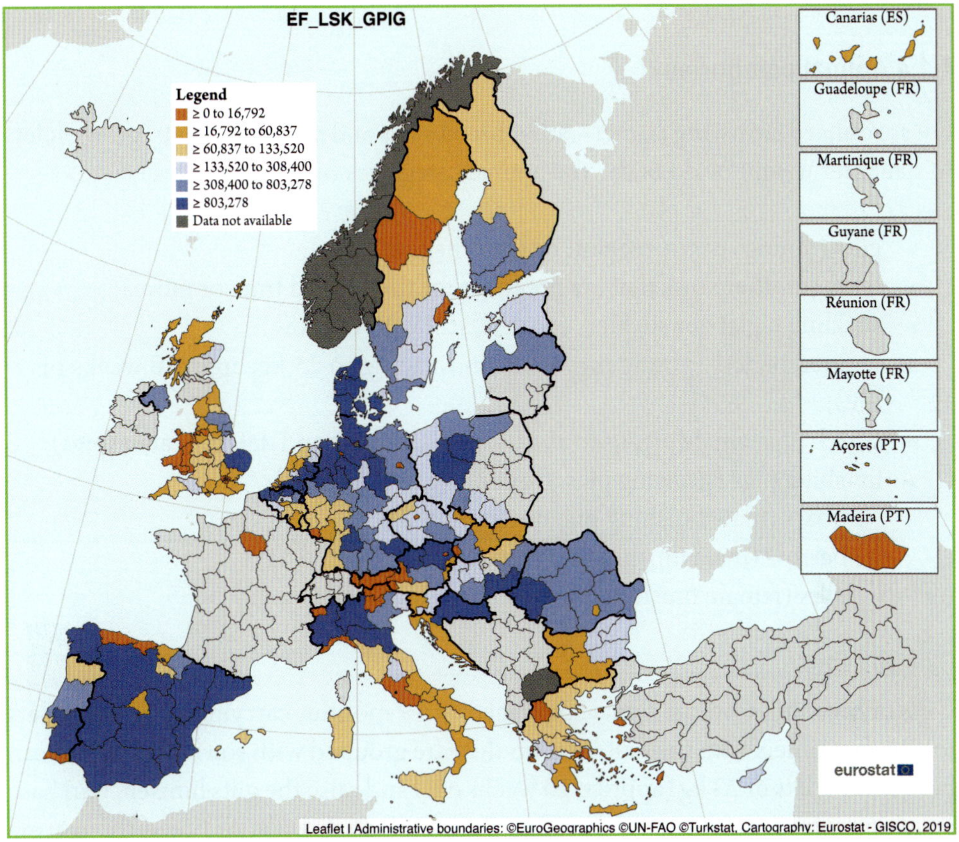

4. Precision pig farming

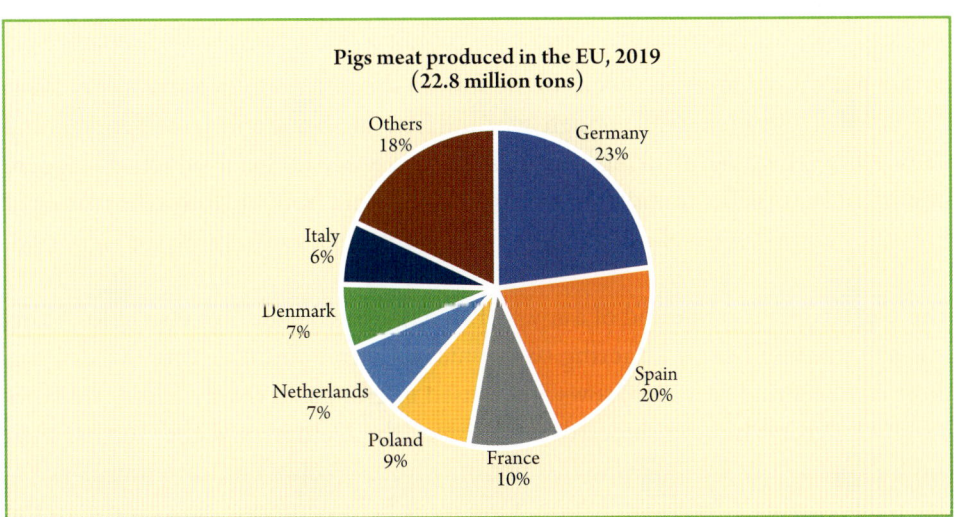

Figure 4.6. Pig population by region (European Union, 2020; http://www.ec.europa.eu).

Figure 4.7. Pig meat production in the EU (modified from European Union, 2020; http://www.ec.europa.eu).

Precision technology and sensor applications

4.1.4 Animal categories

On specialised farms, pig farmers either breed sows and produce piglets, grow piglets to slaughter weight or do both. The animal categories on a pig farm are:

- gilts (young sows, pregnant or with the first litter);
- sows (mother sows that are pregnant for the second time or more);
- suckling piglets or sucklers (piglets, still with the sow);
- weaned piglets (young pigs up from weaning to 25 kg, appr. 10 weeks of age);
- search boar (male pig, used as a boar to stimulate and detect heat in sows);
- finishing pigs (growing pigs, from 25 kg to slaughter):
  - boars (male finishing pig, not castrated);
  - barrow (male finishing pig, castrated);
  - gilts (female finishing pig).

Gilts and sows

Gilts or breeding sows are female pigs aged 6 to 9 months, carrying a litter of piglets for the first time. From the first litter on they are grouped with sows. The farmer can purchase piglets of 23 kg (approx. 10 weeks old) and raise the gilts himself, purchase insemination-ready gilts (6-7 months of age) to use directly, or breed the rearing gilts himself. Sows are female pigs that have already raised a litter of piglets and are inseminated for the second time or up. They are used to produce piglets. A sow will produce piglets that grow into slaughter pigs. For an optimal production of piglets, a sow needs to be in good health and be inseminated at the right age and condition. The farmer must provide for the needs of the sow, so she can reproduce and stay healthy. A healthy environment with enough light and fresh air, being able to perform natural behaviours like eating, drinking, resting and a feed suited for her production stage are important factors. The foundation for a good sow is laid from birth and the rearing of the young animal. A timeline with important events for a sow is shown in Figure 4.8.

The average gestation period of a sow is 115 days (Sasaki and Koketsu, 2007), this is easy to remember by adhering to the maxim 'three months, three weeks and three days'. During gestation, sows are housed in groups. In the EU, pregnant sows have to be group-housed since 2013. However, the Netherlands goes further than the EU regulations by requiring this not only from 28 days but from four days (4-day requirement) after insemination. One week before littering, sows move to the farrowing pens, where they are individually housed. A sow gives birth to 10-22 piglets

4. Precision pig farming

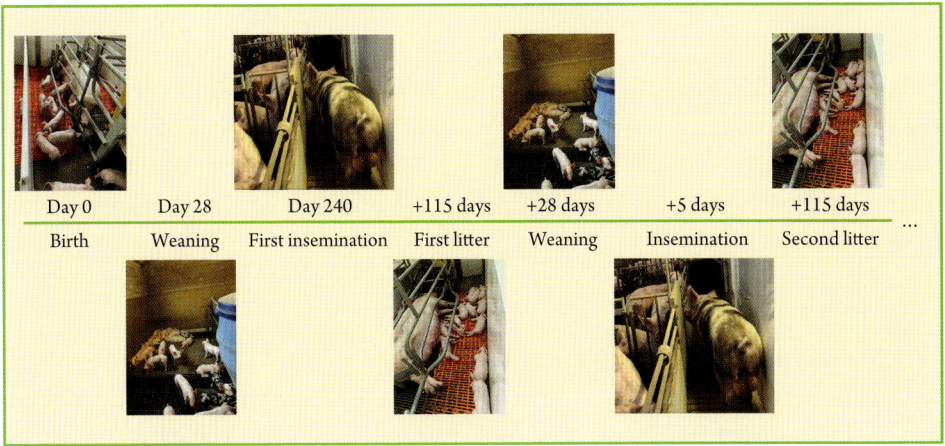

Figure 4.8. Timeline for a sow from birth to first- and second-time littering (photos: HAS University of Applied Sciences).

at a time. Piglets stay with the sow for three to five weeks in the farrowing pen. In a regular farrowing pen, the sow is confined between two metal racks to avoid crushing of piglets, and the piglets can move freely in the pen. For the piglets, there is a heated area where they can rest safe and warm. Sows are fed according to their lactation stage and condition. After the piglets are weaned, the sow goes to the breeding department. There she spends an average of one to two weeks, where she is mated by the boar or via artificial insemination (AI). In the breeding department, sows are housed individually in small pens or crates. Immediately afterwards she goes to the pregnant sow department in group housing. The piglets are moved to weaned pig pens where they are housed in groups of 10-100 piglets, and where they grow until appr. 25 kg or 10 weeks of age. Pig pens or units usually have partially slatted floors, an automatic climate condition system, and automatic feeding systems.

Theoretically, the whole cycle from the start of the pregnancy via birth and lactation to weaning the piglets and being inseminated again can take place within 145 days: 114 days of pregnancy, 26 days of nursing, and 5-6 days empty period. If the sow is not pregnant after insemination or mating, she will usually be in heat again after 21 days. On average, a sow is replaced after 6-7 litters. At this age, the number of piglets per litter decreases, making the sow economically less interesting. Fertility also decreases. The sow becomes pregnant less quickly, which increases the number of loss days, and the number of litters per year (litter index) decreases and the litters will be less uniform.

Boars

Boars are ready for mating from the age of about 8 to 9 months. Most breeders mainly use the boar as a search boar. The boar is led in front of the sows and sniffs out the sows in heat. This is to make insemination as natural as possible: contact with the boar stimulates heat in the sow. During AI, the presence of a boar ensures better fertilisation results.

Finishing pigs

When weaned piglets reach an age of 10 weeks or a weight of 25 kg they are moved to the finishing pens on a closed farm or sold to a finishing farm. Finishing pigs are housed in groups of 10-100 pigs and grow to a weight of approximately 120-130 kg (85-90 kg slaughter weight). Pens have partially slatted floors and a pig density of 0.7 m^2 per pig. An all-in all-out system is used where pens are cleaned and disinfected after each round. Finishing pigs are fed liquid feed or pellets and often by-products are used in the feed; these are the rest products from mostly the food industry.

4.1.5 Main trends in management

Important trends in the management of Dutch pig farms are the employment of extra staff, a growing emphasis on reduction of risk, reduction of production costs, sustainability aspects of pig farming, and the view of society on pig farming.

Mainly due to growing pig numbers per farm, extra staff is employed besides the owner/farmer. This changes the task of a pig entrepreneur considerably. It also leads to more emphasis on using standard operating procedures for every process on the farm, so that the workflow is more equal and a guaranteed minimum standard can be held. In developing and applying precision livestock farming systems, it is important to consider who will work with these systems. The PLF systems should be tuned to the user, and the advice or alerts that it produces should go to the right person, either the owner of the farm or the animal caretaker.

There is a growing emphasis on reducing risks in larger companies. Those risks involve production risks, financial risks, and market risks. There is ongoing attention to the reduction of production costs per kg of pig meat produced. As feed is the most important factor in the cost price of pig meat, much emphasis is on reducing feed costs. Further, labour is an important factor. In a lot of cases, it can be hard to find employees to do the required job, and farmers are looking for ways to make pig farm work more attractive.

There is more attention to the sustainability aspects of the production of pig meat, like CO_2 footprint, feed print, circularity, welfare, environment. Demands on these aspects come from society and consumers. Also, the Dutch government has stated that circular agriculture has the future, so all farmers have to move in this direction. In the same line, society is looking towards pig production with a critical eye. Farmers are considering how to respond to these varying and sometimes negative views.

4.1.6 Main concerns in management

The main concerns in management for Dutch pig farmers are the government programs to reduce the number of pigs and farms, and disease status, pig health and zoonotic diseases. The general trend of Dutch policy is on the reduction of numbers, so pig farmers have to choose where their future possibilities lie and what strategy is most optimal. Government programs to reduce the number of farms and pigs threaten the future possibilities of individual farmers.

The health status of the pig herd is described in 'One health'. One Health is defined as the collaborative efforts of multiple disciplines working locally, nationally, and globally, to attain optimal health for people, animals and our environment, by the One Health Initiative Task Force (OHITF) (American Medical Veterinary Association, 2008).

With the pig industry in the Netherlands being quite concentrated, the risk of outbreaks of zoonoses is high if necessary precautions are not carried out properly. Zoonoses are infectious diseases that can transfer from animals to humans. Well-known examples are rabies in dogs and Q-fever in goats. The Covid-19 pandemic is caused by a Coronavirus that is classified as a zoonotic disease, probably originating from bats, although it is not completely clear how it transferred to humans (Haider *et al.*, 2020). Important pork-borne zoonotic diseases are influenza, *Yersinia*, *Salmonella*, *Trichinella* and *Toxoplasma* (Meemken *et al.*, 2014). Automatic systems for early detection of disease can decrease the risk of outbreaks of zoonoses.

The occurrence of animal diseases that threaten global trade in pig meat, such as African Swine Fever, is a large risk. African Swine Fever (ASF) is one of the most important swine diseases because of severe sanitary and socioeconomic risks. When an infection occurs on the farm, all animals are euthanised on that farm and farms close by. There is no vaccine available for ASF. Wild boars in Europe form a reservoir for infection with ASF, therefore there is always a risk of introduction on pig farms (Sánchez-Vizcaíno et al., 2012).

Pig health is not only a public health issue but first and foremost an important factor in management. Major concerns in the different animal categories are piglet crushing in suckling piglets, tail biting in weaned and finishing pigs and lung problems in weaned and finishing pigs.

Sows give birth to large litters of 12-16 piglets, and of every litter, 1-2 piglets do not survive, due to weakness at birth or to being crushed by the sow. In most housing systems the sows have no opportunity to build a nest to protect the piglets. In the pen, there is a separate, heated area for the piglets, the piglet nest, and the sow lies in the farrowing crate or is loose in the same pen. However, in the first days of life, when resting or sleeping, piglets tend to lie close to the sow instead of in the piglet nest. This poses a risk: when the sow lies down or changes position, piglets can be trapped underneath her and get injured or suffocate. If by chance the farmer notices a piglet getting trapped, he can alert the sow and get her to stand up, but often this is overlooked or the farmer is not there. Restless sows have a higher chance of crushing their piglets, just as inexperienced sows. Farrowing crates are designed to protect the piglets from crushing by confining the sow between bars, but in both confined and loose systems, piglets get crushed (Weber et al., 2007). There is some research into automated systems to decrease piglet crushing (Manteuffel et al., 2017; Oczak et al., 2015).

Tail- and ear-biting are multifactorial behavioural problems that can occur in weaning pigs and finishing pigs but can start already in the farrowing pen. Factors that increase the risk of tail- and ear-biting are stress, crowding, climate conditions (draught, ammonia), diet (amount of protein) and genetic factors. Tail and ear-biting can lead to wounds and infections, and in severe cases also flank biting and leg biting can occur. In the slaughterhouse, carcasses with severe bite wounds will be rejected and the farmer receives a financial penalty (Breuer et al., 2005; Van Der Meer et al., 2017). Early, automated detection of the behaviours at the onset of tail biting, or controlling the factors that increase the risk of tail biting, such as climate conditions, can decrease the risk. An early detection system for tail biting using image recognition had been studied (D'Eath et al., 2018).

A major health concern in weaned and finishing pigs is the occurrence of lung problems. An important risk factor is the climate in the pen. Young pigs are sensitive to draught, and high ammonia levels increase the sensitivity of the respiratory systems for infections. Lung problems lead to decreased welfare, slower growth rate and a penalty in the slaughterhouse when lung lesions are found in the carcass (Michiels *et al.*, 2017). Climate control systems can help the farmer prevent lung problems in the pigs, and an early warning system for coughing can help to prevent severe infections (Silva *et al.*, 2008).

4.2 Application of Precision Livestock Farming in pig farming

In PLF, wearable technologies dominate the market. However, in less-value-per-animal systems like sheep, goat, pig, poultry, and fish, one sensor, like a camera or robot per herd/flock/school, rather than one sensor per animal, will become common (Halachmi *et al.*, 2019). Pigs are housed in groups, ranging from 10 to >100 individuals, which makes monitoring a challenge. For growing and finishing pigs, most PLF systems are aimed at group monitoring, while for sows a few systems are aimed at individual monitoring. If individual recognition for pigs becomes more common, the number of individual sensor systems will also increase. At the moment, individual sensors for pigs are still too expensive, and there is a technical challenge in recognising individuals. Group monitoring, however, is becoming easier with advanced software techniques, such as machine learning and deep learning, and if the massive amounts of data that are acquired are being translated into information, these systems will greatly help farmers and farming in improving pig welfare and production (Benjamin and Yik, 2019; Halachmi *et al.*, 2019).

4.2.1 Dashboards

Many sensors gather data in the pig farm, but to use this data in management, it has to be translated into information for the farmer. This is where dashboards are used. Several companies have developed dashboards for farmers, where farm data and sometimes also sensor data are visualised, and analyses are shown. Farmers can use these dashboards to plan and work on the company strategy.

The first dashboards were developed with the purpose to visualise on-farm data for the farmer, to aid the farmer in the day to day running of the farm. Information on climate conditions and feed intake were visualised for the farmer to be able to detect anomalies and, if necessary, adjust the ventilation or the feed. The next-generation dashboards are also showing analyses of the data, for example, the water to feed ratio or real and predicted growth rates. Software techniques such as machine learning can be used to find these anomalies or calculate the predictions. Many companies offer commercial dashboards for pig farmers (Table 4.1).

In pig farming, dashboards are also used for strategic planning. In interactive dashboards, farmers can compare results between different sow groups or see the results of breeding strategies, using the results or the advice to make management decisions.

4.2.2 Group or batch monitoring

Many systems in pig farming are aimed at monitoring pig groups or units. Especially for weaned piglets and finishing pigs, almost all acquired data are group totals or averages such as feed or water intake per unit, average growth rate or coughing rate per pig unit. Since animals are not individually identified, no individual data can be gathered. Health and production can only be monitored per group and not per animal – diagnosis and treatment consequently are also per group. However, detecting disease signals such as an increased number of coughs or a deviating behaviour in an early stage can limit the number of pigs that have to be treated, or prevent the spread of disease through the farm.

Table 4.1. an overview of dashboards for pig management.

Company	Dashboard	Website
Agrisyst	PigExpert	www.agrisyst.com
FarmResult	PigResult	www.farmresult.com
Agrovision	PigVision	www.agrovision.com
Hotraco	Prisma Farm Management	www.Hotraco-Agri.com
Plan-a-head	Pig Software	www.planaheadgroup.com
Claas	Pig farm management	www.claas-e-systems.com
Garth Pig Practice Ltd	Pig Focus	www.pigfocus.com
Herdstar	GF Pro	www.herdstar.com
Prüllage Systeme	Smart Farming software	www.pruellage.de
Agritec	Porcitec	www.agritecsoft.com
Folio3 software	Swine record keeping	www.animalcare.folio3.com
Isagri	Swine management system	www.isagri.com
Agrosoft	WinPig.NET	www.agrosoft.eu

Behaviour: cameras

Using image analysis, location, distribution and activity of pigs can be detected in groups of pigs. In a Swedish study, nine pigs in a pen were filmed with a top view camera and images were analysed to determine the proportion of pigs in different pen areas. Changed behaviour and location of the pigs could indicate changed climate conditions or health issues (Nilsson *et al.*, 2015).

Weight: cameras

Image analysis can be used to estimate weight, monitoring the average weight of a group of pigs without identifying individuals (Kongsro, 2014; Marchant *et al.*, 1999; Pezzuolo *et al.*, 2018). The EyeGrow camera system of Fancom estimates the weight of finishing pigs that come into view of the camera (https://tinyurl.com/nfz3nrwy). Practical issues with camera systems in pig farms are lighting and background: dark-coloured or dirty pigs are difficult to distinguish from a dark background (i.e. a dark floor). Infrared cameras can be used at night or in dark circumstances (Benjamin and Yik, 2019).

Sound: microphones

Lung problems are a frequent problem in finishing pigs, and coughing is the result. Coughing can be detected with microphones, and algorithms can detect an increase in the cumulative number of coughs in a group of pigs (Chedad *et al.*, 2001; Exadaktylos *et al.*, 2008; Silva *et al.*, 2008). The difference between a dry cough and a wet cough (= severe disease) can even be detected by the software, that is commercially available. The cough monitor of Soundtalks detects coughing events in pig units (Figure 4.9). Pig farmers can use this device to detect lung problems at an early stage and decrease the use of antibiotics (Silva *et al.*, 2008). In a German study, sound analysis was used to detect piglets in danger of being crushed by the sow. In a farrowing compartment with four sows, stress sounds of piglets related to trapping were outnumbered by other stress-related sounds of the piglets. However, when the researchers included context parameters, such as the age of the piglets and the body posture history of the sow, the specificity and sensitivity could be increased to 95 and 70%, respectively (Manteuffel *et al.*, 2017).

Figure 4.9. Cough monitor for fattening pigs (photo: Courtesy Soundtalks).

4.2.3 Individual monitoring

Identification

The first step in gathering individual animal data and monitoring health and production on an animal level is identifying individual pigs. This can be done using Radio Frequency Identification (RFID) tags on the pigs, or with several types of image recognition.

RFID chips can be incorporated in an ear tag and can be used to identify pigs. There are two types of RFID chips: low frequency or LF-RFID and ultra-high-frequency or UHF-RFID chips (Benjamin and Yik, 2019). LF-RFID tags can be used to register behaviour and health, and are applied mainly to register feeding or drinking patterns of individual pigs (Maselyne *et al.*, 2014, 2017; Matthews *et al.*, 2017). An RFID system requires an RFID transponder (ear tag) and an RFID antenna or receiver (located at the feeder or drinker). The responder circuit of the tag communicates with a low, high or ultra-high frequency radio wave. When an RFID tag comes within proximity of an RFID reader, it receives a signal. Then a second radio frequency signal carries the data to the reader. These data usually consist of animal and farm information and can be used to identify the animal or the data can be stored. Low-frequency RFID cannot identify individuals when multiple transponders are very close together (such as in one box), but when mounted on pigs they can be used to individually identify pigs, for example when moving as a group through a narrow walkway (M. Cox, personal communication). LF RFID chips have a low range (<1 m). UHF-RFID tags can be used to identify multiple animals at a greater range (3-10 m) but are sensitive to interference, leading to false registrations (Benjamin and Yik, 2019; Matthews et al., 2017). Disadvantages of using ear tags for pigs are loss of tags, pain and stress during tagging, and the need to remove the ear tag in the slaughter process (Benjamin and Yik, 2019).

Optical character recognition is a low-cost system that recognises text characters, such as license plates or QR codes. With these systems, characters or symbols on the ear tag can be read automatically. For this system, a digital camera is needed and machine learning algorithms have been developed to perform the remote identification (Lancaster et al., 2018). The system can also be used to identify animals from written or painted characters on the animal, but these visual patterns tend to disappear or blur from the animals quickly (Benjamin and Yik, 2019).

Facial recognition has been developed for human identification and can now also be used for individual recognition of pigs, with good results: up to 98% recognition can be achieved with high speed (620 images/second). Usually, the snout, top of the head and eye regions are used for recognition (Benjamin and Yik, 2019). In a UK study with ten pigs, 1,553 images of pig faces were used to train a neural network, resulting in an individual pig recognition with an accuracy of 96.7% (Hansen et al., 2018).

Behaviour: cameras and light barriers

Using image analysis, location, distribution and activity of pigs can be detected and combined with automatic identification, individual behavioural patterns can be recognised (Marchant et al., 1999). Image analysis can be used to estimate the weight of individual pigs, either with an app on a tablet or with a fixed camera system. In a study in Belgium, 40 piglets were filmed with a top view camera and from the images, the individual animals were detected and tracked. Locomotion of the piglets could be determined with an accuracy of 89.8% (Kashiha et al., 2014). Aggressive behaviours, walking patterns and postures of (lactating) sows can be detected. Using image recognition, aggression could be detected in weaned piglets (Viazzi et al., 2014), and during aggressive interactions in 14-week old pigs, facial expressions could be used to detect intent and emotion in the animals (Camerlink et al., 2018). In a German study, light barriers were used to measure the activity of 34 sows that were near parturition.

Based on the activity, the onset of parturition could be predicted with a sensitivity of 88% and alerts could be given 13-24 hours before parturition with an accuracy of ±4 hours (Manteuffel *et al.*, 2015).

Behaviour: accelerometers

In sows, accelerometers can be used to monitor activity and posture (e.g. lying, walking) (Escalante *et al.*, 2013; Liu *et al.*, 2018). Although in dairy cows these type of sensors are widely used, this is not the case in pig farming. It is possible to detect the increased activity of sows in heat with accelerometers, but since sows come in heat very regularly and at approximately the same time within a group of breeding sows, this device would not increase heat detection very much – oestrus detection is already very high in pig production, so the added value of such a system is small. Research is being done into using accelerometers to detect lameness in sows; in group-housed sows, this can be quite a problem and it is not easy to spot for the farmer. Using ear tags, lameness could be detected with an accuracy of 0.57-0.93 (Scheel *et al.*, 2017; Traulsen *et al.*, 2016). In a study in Vienna, nest-building behaviour of nine loose housed sows was recorded with accelerometers. It was possible to classify nest-building events with a sensitivity of 87% and a specificity of 85%. With such a detection system, crating of sows could be postponed until the parturition starts and limited to the first few days after farrowing (Oczak *et al.*, 2015).

Temperature sensors and infrared imaging

Piglets have no insulation or body fat, therefore their skin temperature directly reflects their core temperature; in older pigs, warm circumstances will cause their skin temperature to rise while cold circumstances will result in lower skin temperatures. Temperature monitoring could be used as a detection system for disease or stress situations. When a pig is sick and has a fever, this increase in temperature could be detected; stress situations also influence body temperature, where core temperature usually increases and temperature in the extremities decreases. Temperature sensors can be used on the body of a pig, where it has to have contact with the skin, for example in an ear tag (Andersen *et al.*, 2008). Remote measuring using infrared imaging with a thermographic camera is also possible (Nääs *et al.*, 2014). In slaughter pigs, correlations were found between ear temperature and cortisol levels in the blood, suggesting that stress levels can be measured with a thermometer (Warriss *et al.*, 2006). No temperature monitoring systems are yet commercially available for pigs.

Weighing devices

Weight estimations using cameras is not yet commercially available for individual animals, but adding an RFID reader to a weight estimation camera would solve this quickly. Automatic weighing in pigs is possible by placing a weighing device in the pen and automatically recording the weights of the animals (Williams *et al.*, 1996). Electronic feeders combined with weighing scales can be used to sort finishing pigs into weight groups with different diets, or to adjust the diet for the pregnant sow (Murphy and De Lange, 2004; Zimmerman *et al.*, 2004). The use of these or similar feeding and sorting techniques that allow the proper amount of feed with the suitable composition to be supplied on time to a group of animals or individual animals is called precision feeding (Pomar and Remus, 2019).

Automatic syringe

Giving vaccinations in pigs are processes that involve manual work and an administration process. To ease this process for the farmer, automatic syringes have been developed. With these syringes, injections can be given automatically so that it is less exhausting, the injected volume is standardised and temperature-controlled vaccination is possible. Parameters, such as injection volume are set centrally and monitored in each device, and alerts are given when the bottle or the needle needs to be replaced. These devices send and receive data over WiFi and are controlled by a tablet or smartphone. Information on the number of injections, missed injections, the volume of the vaccine in the bottle and vaccine temperature are sent (Health Europa, 2019).

4.2.4 Summary of PLF systems for group and individual monitoring

A summary of PLF systems for group monitoring and individual monitoring of pigs is given in Table 4.2. Mentioned manufacturers are examples and not a complete overview.

Table 4.2. Summary of precision livestock farming sensors for pigs, with examples of manufacturers.

PLF sensor	Where	What it measures	Why	Commercial product	Company website
Accelerometer	ear tag	activity	nest building, health, farrowing, oestrus	yes	RemoteInsights.net
Sound analysis	microphone in pen	coughing	health	yes	Soundtalks.com
Sound analysis	microphone in pen	stress calls	stress	no	-
Vision	topview camera	bodyweight	production, health	yes	Fancom.com
Vision	camera	behaviour	heat stress, thermal comfort	no	-
Vision	tablet, smartphone	bodyweight	production	yes	Itochu.co.jp
Vision	camera	locomotion	lameness	no	-
Vision	camera	tail posture	tail biting	no	-
Vision	topview camera	behaviour	health status	yes	Serket-tech.com
Vision	camera	location	health, heat stress	no	-
Vision	camera	behaviour	aggression	no	-
Vision	camera	comfort behaviour, posture	health, climate conditions	no	-
Vision	camera	static and moving images	climate monitoring	yes	www.iconize.com
Vision	camera	face recognition	identification	yes	m.yingzi.com
Vision and climate	topview camera and device	behaviour and climate	health	yes	Healthyclimate.nl
Climate and smell	pig unit	climate, smell	health and production	yes	Slimmestal.nl
Weighing device	electronic sow feeder	weight, feed intake	production, health, sorting	yes	www.nedap.com
Weighing device	electronic feeder	weight, feed intake	production, sorting	yes	www.nedap.com
RFID	RFID	presence at boar station	heat detection	yes	www.nedap.com
Light barriers	farrowing crate	activity	parturition	no	-
Several sensors	syringe	vaccinations	health	yes	syrinjector.com

4.3 Monitoring the technical systems

In pig farms, monitoring the technical systems might be as important as monitoring the animals. If the ventilation system breaks down, this can have immediate and severe consequences for the animals and the farm. Animals may overheat, become ill or even die, leading to animal suffering and economic losses. Deferred maintenance can lead to disasters, such as fire, but also less acute problems, such as irregular feed distribution or suboptimal climate conditions. These problems can be detected by monitoring the systems or the animals, but prevention or even prediction of the problems is better. Predictive maintenance is an upcoming field of research and very interesting for agriculture (Lüttenberg et al., 2018).

4.3.1 Preventive maintenance and predictive maintenance

There are three ways of maintaining the farm systems, such as the indoor climate system (fans, heating), feeding system and water system. One way is to wait until the system breaks down before doing repairs. If you compare it with car maintenance, this would be driving until your car breaks down. Hopefully, you can get to a garage but in the worst scenario, you have to be towed away. A better way is to plan maintenance, regularly check the systems and replace parts of the system that are prone to damage, just like you have your regular service checks at your garage, every year or every 15,000 kilometres. This is called preventive maintenance. A third way is to use sensors to signal damage or wear of machine parts in an early stage and use that information to perform maintenance. This is called predictive maintenance. Sensors can be used to detect damage or wear by measuring (increased) voltage, sound, vibration and machine learning techniques are used to predict the chance of system failure by calculating the chance of exceeding a certain threshold (Lee et al., 2019). A predictive car maintenance strategy can reduce the number of costly incidents, by suggesting the driver to schedule a visit to the dealer, once the failure probability within a certain period exceeds a pre-defined threshold. The condition of each subsystem in a car can be monitored with onboard vehicle telematics systems, which become increasingly available in modern cars (Last et al., 2010). Some predictive maintenance systems for cars even take driving style into account: a rough driving style will lead to different maintenance advice (Chowdhury et al., 2015). In livestock farming, several mechanical systems can be outfitted with sensors to predict their maintenance. The most important disaster to prevent is fire, which often starts in the fuse box. A heat sensor located there can send out a warning at a very early stage so that an incident can be prevented. A malfunctioning fan in a unit with finishing pigs on a warm day can also be detrimental, with animals overheating quickly. There are fans with built-in sensors that will monitor the functionality of the fan and that will alert the farmer in time for maintenance and to prevent malfunctioning (www.fancom.com).

4.3.2 Monitoring group behaviour to detect events

Systems that monitor animal behaviour will show changes in the conditions around the animals because the animals will change their behaviour according to the changes around them. For example, if there is a draught, pigs will be more restless and change their lying postures; if it is cold, pigs will lie close together, and when it is warm, they will spread out. Changes in the climate system can therefore be noticed by a system that monitors behaviour. An example is the camera system of EyeNamic, originally developed for broilers, that monitors distribution and activity. If distribution or activity is deviating from the expected values, the farmer gets an alert (Kashiha *et al.*, 2014; Leroy *et al.*, 2006).

4.4 Future trends in precision pig farming

4.4.1 Dashboards and data integration

Dashboards that visualise and analyse data and give the farmer information of his operation, make it possible to make data-driven decisions based on a combination of sensor and management data. An example is a cooperation between MS Schippers and Agrisyst, they developed a system that combines weighing data from the PigScale with feed data from the feeding computer. This system, the FeedConversionMonitor, gives a weekly overview of combined weighing and feed data and daily feed conversion rates. This allows the farmer to use the information on feed intake, growth, feed conversion ratio and costs, and to determine which pigs give the best return on investment. PigResult (FarmResult) is another example of a management system developed for the grower to finisher stage in pigs. In a smart dashboard, all aggregated data can be overviewed, including automatically generated process data (real-time), manual farm data and supply chain data. Analysing all data coming in to and going out of the farm helps to visualise practical solutions to problems as well as creating biostatistics to improve performance.

4.4.2 Vision and machine learning

Several research groups and companies are working on automatically analysing vision data and detecting anomalies such as health issues or problems with climate conditions or feed and water installations. An example of health monitoring based on machine learning is the system of Serket, which analyses images of finishing pigs to automatically detect health problems. The systems use machine learning to train the algorithms and labelled data are used, with the farmer giving input as to what is happening in the pens and what issues he detected concerning the pigs. This system is still in development, but a promising way of monitoring pigs and alerting the farmer in an early stage.

4.4.3 Sound

The cough monitor is paving the way for other health or production monitoring systems based on sound. Several studies are being done into sound analysis. Two pathways can be used: either the sound patterns in the farm are monitored, and anomalies in the patterns are detected, or specific sounds are detected. When monitoring patterns, the farmer will be alerted when for example the feeding system is malfunctioning, or a fan breaks down: all animals will show different behaviour and the sound pattern will change in that pig unit. In the case of detecting specific sounds, the algorithm will detect for example coughing and sneezing sounds such as in the cough monitor, or screams. If the number of these sounds exceeds a threshold, the farmer can be alerted that there is a problem with sick animals (coughing) or aggression (screaming). In pig farms, sound monitoring is difficult because of the background noise. Filters have to be applied to detect specific sounds. Monitoring of patterns is easier, but even then the algorithms have to take into account the background noise of fans and feeding equipment.

4.4.4 Practical applications

In the pig sector, as in most sectors, sensor and data applications will only be implemented if they are practical, saving time and easing the farm management process. Many sensor and data applications are being developed and tested in non-commercial circumstances, such as on research farms. However, only robust devices, and practical, labour-extensive systems that can be applied in the everyday practice of pig farming, are fit for the future. We see that in nucleus farms, at the top of the breeding pyramid, integrated PLF systems are implemented. Those farmers, who are focused on gathering data but also have to work efficiently, will use systems that save time and money. An example is automated weighing with RFID identification of sows. Instead of manually weighing sows and recording weights by hand, an automated weighing system can scan the sow's RFID and automatically store her weight. This saves time and money. Those types of systems will probably first be implemented in top breeding farms that work for breeding companies, and then flow to commercial farms.

References

American Medical Veterinary Association, 2008. One Health: a new professional imperative, One Health Initiative Task Force Final Report.

Andersen, H.M.L., Jørgensen, E., Dybkjær, L., Jørgensen, B., 2008. The ear skin temperature as an indicator of the thermal comfort of pigs. Applied Animal Behaviour Science 113: 43-56. https://doi.org/10.1016/j.applanim.2007.11.003

Benjamin, M., Yik, S., 2019. Precision livestock farming in swine welfare: a review for swine practitioners. Animals 9: 133. https://doi.org/10.3390/ani9040133

Breuer, K., Sutcliffe, M.E.M., Mercer, J.T., Rance, K.A., O'Connell, N.E., Sneddon, I.A., Edwards, S.A., 2005. Heritability of clinical tail-biting and its relation to performance traits. Livestock Production Science 93: 87-94. https://doi.org/10.1016/j.livprodsci.2004.11.009

Camerlink, I., Coulange, E., Farish, M., Baxter, E.M., Turner, S.P., 2018. Facial expression as a potential measure of both intent and emotion. Scientific Reports 8: 17602. https://doi.org/10.1038/s41598-018-35905-3

CBS, 2020. Statline: landbouw: gewassen, dieren, grondgebruik en arbeid op nationaal niveau. Available at: https://tinyurl.com/tjc658ue.

Chedad, A., Moshou, D., Aerts, J.M., Van Hirtum, A., Ramon, H., Berckmans, D., 2001. Recognition system for pig cough based on probabilistic neural networks. Journal of Agricultural and Engineering Research 79: 449-457. https://doi.org/10.1006/jaer.2001.0719

Chowdhury, A., Banerjee, T., Chakravarty, T., Balamuralidhar, P., 2015. Smartphone based sensing enables automated vehicle prognosis. In: 9th International Conference on Sensing Technology (ICST), pp. 452-455. https://doi.org/10.1109/ICSensT.2015.7438441

D'Eath, R.B., Jack, M., Futro, A., Talbot, D., Zhu, Q., Barclay, D., Baxter, E.M., 2018. Automatic early warning of tail biting in pigs: 3D cameras can detect lowered tail posture before an outbreak. PLoS ONE 13: e0194524. https://doi.org/10.1371/journal.pone.0194524

Escalante, H.J., Rodriguez, S. V., Cordero, J., Kristensen, A.R., Cornou, C., 2013. Sow-activity classification from acceleration patterns: a machine learning approach. Computers and Electronics in Agriculture 93: 17-26. https://doi.org/10.1016/j.compag.2013.01.003

European Commission, 2020. Agricultural Census 2020, Eurostat. Available at: https://ec.europa.eu/eurostat/web/agriculture/census-2020.

European Union, 2020. Agriculture, forestry and fishery statistics, Eurostat. Publication Office of the European Union, Luxembourg, Luxembourg. Available at: https://tinyurl.com/9d6un96n.

Exadaktylos, V., Silva, M., Aerts, J.M., Taylor, C.J., Berckmans, D., 2008. Real-time recognition of sick pig cough sounds. Computers and Electronics in Agriculture 63: 207-214. https://doi.org/10.1016/j.compag.2008.02.010

Haider, N., Rothman-Ostrow, P., Osman, A.Y., Arruda, L.B., Macfarlane-Berry, L., Elton, L., Thomason, M.J., Yeboah-Manu, D., Ansumana, R., Kapata, N., Mboera, L., Rushton, J., McHugh, T.D., Heymann, D.L., Zumla, A., Kock, R.A., 2020. COVID-19 – Zoonosis or emerging infectious disease? Frontiers in Public Health 8: 763. https://doi.org/10.3389/fpubh.2020.596944

Halachmi, I., Guarino, M., Bewley, J., Pastell, M., 2019. Smart animal agriculture: application of real-time sensors to improve animal well-being and production. Annual Review of Animal Biosciences 7: 403-425. https://doi.org/10.1146/annurev-animal-020518-114851

Hansen, M.F., Smith, M.L., Smith, L.N., Salter, M.G., Baxter, E.M., Farish, M., Grieve, B., 2018. Towards on-farm pig face recognition using convolutional neural networks. Computers in Industry 98: 145-152. https://doi.org/10.1016/j.compind.2018.02.016

Health Europa, 2019. SyrinJector takes livestock vaccination to the next level. Health Europa Quarterly 8: 194-197.

Kashiha, M.A., Bahr, C., Ott, S., Moons, C.P.H., Niewold, T.A., Tuyttens, F., Berckmans, D., 2014. Automatic monitoring of pig locomotion using image analysis. Livestock Science 159: 141-148. https://doi.org/10.1016/j.livsci.2013.11.007

Kongsro, J., 2014. Estimation of pig weight using a Microsoft Kinect prototype imaging system. Computers and Electronics in Agriculture 109: 32-35. https://doi.org/https://doi.org/10.1016/j.compag.2014.08.008

Lancaster, J.M., Psota, E., Mote, B.E., Perez, L., Fricke, L., Mittek, M., Kett, L.E., Schmidt, T.B., 2018. Evaluation of a novel computer vision systems' ability to continuously identify and track the activities of newly weaned pigs. Journal of Animal Science 96 (suppl 2): 69-70. https://doi.org/10.1093/jas/sky073.129

Last, M., Sinaiski, A., Subramania, H.S., 2010. Predictive maintenance with multi-target classification models. Asian Conference on Intelligent Information and Database Systemss. pp. 368-377.

Lee, S.M., Lee, D., Kim, Y.S., 2019. The quality management ecosystem for predictive maintenance in the Industry 4.0 era. International Journal of Quality Innovation 5: 4. https://doi.org/10.1186/s40887-019-0029-5

Leroy, T., Mentasi, T., Costa, A., Guarino, M., Aerts, J.M., 2006. Real-time measurement of pig activity in practical conditions. In: Proceedings of the Fourth Workshop on Smart Sensors in Livestock Monitoring, pp. 12-14.

Liu, L.S., Ni, J.Q., Zhao, R.Q., Shen, M.X., He, C.L., Lu, M.Z., 2018. Design and test of a low-power acceleration sensor with bluetooth low energy on ear tags for sow behaviour monitoring. Biosystems Engineering 176: 162-171. https://doi.org/10.1016/j.biosystemseng.2018.10.011

Lüttenberg, H., Bartelheimer, C., Beveringen, D., 2018. Designing predictive maintenance for agricultural machines. In: Twenty-Sixth European Conference of Information Systems (ECIS2018), Portsmouth, UK, pp. 1-17.

Manteuffel, C., Hartung, E., Schmidt, M., Hoffmann, G., Schön, P.C., 2015. Towards qualitative and quantitative prediction and detection of parturition onset in sows using light barriers. Computers and Electronics in Agriculture 116: 201-210. https://doi.org/10.1016/j.compag.2015.06.017

Manteuffel, C., Hartung, E., Schmidt, M., Hoffmann, G., Schön, P.C., 2017. Online detection and localisation of piglet crushing using vocalisation analysis and context data. Computers and Electronics in Agriculture 135: 108-114. https://doi.org/10.1016/j.compag.2016.12.017

Marchant, J.A., Schofield, C.P., White, R.P., 1999. Pig growth and conformation monitoring using image analysis. Animal Science 68: 141-150. https://doi.org/10.1017/S1357729800050165

Maselyne, J., Van Nuffel, A., Briene, P., Vangeyte, J., De Ketelaere, B., Millet, S., Van Den Hof, J., Maes, D., Saeys, W., 2017. First results of a warning system for individual fattening pigs based on their feeding pattern. In: Precision Livestock Farming 2017 – Papers Presented at the 8th European Conference on Precision Livestock Farming, ECPLF 2017. European Conference on Precision Livestock Farming, pp. 503-511.

Maselyne, J., Van Nuffel, A., De Ketelaere, B., Vangeyte, J., Hessel, E.F., Sonck, B., Saeys, W., 2014. Range measurements of a high frequency radio frequency identification (HF RFID) system for registering feeding patterns of growing-finishing pigs. Computers and Electronics in Agriculture 108: 209-220. https://doi.org/10.1016/j.compag.2014.08.006

Matthews, S.G., Miller, A.L., Clapp, J., Plötz, T., Kyriazakis, I., 2017. Early detection of health and welfare compromises through automated detection of behavioural changes in pigs. The Veterinary Journal 217: 43-51. https://doi.org/10.1016/j.tvjl.2016.09.005

Meemken, D., Tangemann, A.H., Meermeier, D., Gundlach, S., Mischok, D., Greiner, M., Klein, G., Blaha, T., 2014. Establishment of serological herd profiles for zoonoses and production diseases in pigs by 'meat juice multi-serology'. Preventive Veterinary Medicine 113: 589-598. https://doi.org/10.1016/j.prevetmed.2013.12.006

Meerburg, B.G., Pest, D., 2014. It's good to be sure : product development & risk management in the animal breeding industry. Nyenrode Business University, Breukelen, the Netherlands.

Michiels, A., Vranckx, K., Piepers, S., Del Pozo Sacristán, R., Arsenakis, I., Boyen, F., Haesebrouck, F., Maes, D., 2017. Impact of diversity of *Mycoplasma hyopneumoniae* strains on lung lesions in slaughter pigs. Veterinary Research 48: 2. https://doi.org/10.1186/s13567-016-0408-z

Murphy, J., De Lange, K., 2004. Nutritional strategies to minimize nutrient output. In: Murphy, J.M., Kane, T.M., De Lange, C.F.M. (eds.) Proceedings of the 2004 London Swine Conference. London Swine Conference, London, Ontario, Canada, 170.

Nääs, I.A., Garcia, R.G., Caldara, F.R., 2014. Infrared thermal image for assessing animal health and welfare. Journal of Animal Behaviour and Biometeorology 2: 66-72. https://doi.org/10.14269/2318-1265/jabb.v2n3p66-72

Nilsson, M., Herlin, A.H., Ardö, H., Guzhva, O., Aström, K., Bergsten, C., 2015. Development of automatic surveillance of animal behaviour and welfare using image analysis and machine learned segmentation technique. Animal 9: 1859-1865. https://doi.org/10.1017/S1751731115001342

Oczak, M., Maschat, K., Berckmans, D., Vranken, E., Baumgartner, J., 2015. Classification of nest-building behaviour in sows on the basis of accelerometer data. Biosystems Engineering 140: 48-58. https://doi.org/10.1016/j.biosystemseng.2015.09.007

Pezzuolo, A., Guarino, M., Sartori, L., González, L.A., Marinello, F., 2018. On-barn pig weight estimation based on body measurements by a Kinect v1 depth camera. Computers and Electronics in Agriculture 148: 29-36. https://doi.org/10.1016/j.compag.2018.03.003

Pomar, C., Remus, A., 2019. Precision pig feeding: a breakthrough toward sustainability. Animal Frontiers 9: 52-59. https://doi.org/10.1093/af/vfz006

Sánchez-Vizcaíno, J.M., Mur, L., Martínez-López, B., 2012. African swine fever: an epidemiological update. Transboundary and Emerging Diseases 59: 27-35. https://doi.org/10.1111/j.1865-1682.2011.01293.x

Sasaki, Y., Koketsu, Y., 2007. Variability and repeatability in gestation length related to litter performance in female pigs on commercial farms. Theriogenology 68: 123-127. https://doi.org/10.1016/j.theriogenology.2007.04.021

Scheel, C., Traulsen, I., Auer, W., Müller, K., Stamer, J., Krieter, E., 2017. Detecting lameness in sows from ear tag-sampled acceleration data using wavelets. Animal 11: 2076-2083.

Silva, M., Ferrari, S., Costa, A., Aerts, J.M., Guarino, M., Berckmans, D., 2008. Cough localization for the detection of respiratory diseases in pig houses. Computers and Electronics in Agriculture 64: 286-292. https://doi.org/10.1016/j.compag.2008.05.024

Traulsen, I., Breitenberger, S., Auer, W., Stamer, E., Müller, K., Krieter, J., 2016. Automatic detection of lameness in gestating group-housed sows using positioning and acceleration measurements. Animal 10: 970-977. https://doi.org/10.1017/S175173111500302X

Van Der Meer, Y., Gerrits, W.J.J., Jansman, A.J.M., Kemp, B., Bolhuis, J.E., 2017. A link between damaging behaviour in pigs, sanitary conditions, and dietary protein and amino acid supply. PLoS ONE 12: e0174688. https://doi.org/10.1371/journal.pone.0174688

Van der Meulen, H., 2020a. Agrimatie. Varkenshouderij. Available at: www.agrimatie.nl.

Van der Meulen, H.A.B., 2020b. Who feeds the Netherlands [Wie voedt Nederland], Voedsel-Economisch Bericht. Available at: https://tinyurl.com/59r6sr7r.

Van der Peet, G., Leenstra, F., Vermeij, I., Bondt, N., Puister, L., Van Os, J., 2018. Feiten en cijfers over de Nederlandse veehouderijsectoren 2018. Wageningen, the Netherlands. https://doi.org/10.18174/464128

Viazzi, S., Ismayilova, G., Oczak, M., Sonoda, L.T., Fels, M., Guarino, M., Vranken, E., Hartung, J., Bahr, C., Berckmans, D., 2014. Image feature extraction for classification of aggressive interactions among pigs. Computers and Electronics in Agriculture 104: 57-62. https://doi.org/10.1016/j.compag.2014.03.010

Warriss, P.D., Pope, S.J., Brown, S.N., Wilkins, L.J., Knowles, G., 2006. Estimating the body temperature of groups of pigs by thermal imaging. VetRecord 158: 331-334.

Weber, R., Keil, N.M., Horat, R., 2007. Piglet mortality on farms using farrowing systems with or without crates. Animal Welfare 16: 277-279.

Williams, S.R.O., Moore, G.A., Currie, E., 1996. Automatic weighing of pigs fed ad Libitum. Journal of Agricultural and Engineering Research 64: 1-10. https://doi.org/10.1006/jaer.1996.0040

Zimmerman, R.G., Sauder, H.H., Zommerman, G.M., 2004. Livestock weighing and sorting apparatus. United States Patent 6805078-B2. Available at: https://www.freepatentsonline.com/6805078.pdf.

5. Sensors and techniques to monitor and improve welfare and performance in poultry chains

A. Lourens[1*], J.L.T. Heerkens[1] and L. Star[1,2]

[1]*Aeres University of Applied Sciences, Dronten, the Netherlands; s.lourens@aeres.nl*
[2]*Schothorst Feed Research, Lelystad, the Netherlands*

Highlights

- The poultry sector is segregated into either meat- or egg-producing farms, and both sectors have highly developed breeding companies that supply genetic material to rearing farms, breeder farms and hatcheries to deliver the final crossbreeds.
- Due to an increasing demand for safe, cheap, healthy, and nutritious protein sources, the poultry sector increases annually on a global scale.
- The general trend is that more animals are housed at fewer farms and under less labour-intensive management.
- The development of smart farming techniques to provide extra eyes on the animals is urgent in this sector to assure sufficiently high animal welfare, health, and production performance levels.
- Smart farming techniques are therefore urgently required to support these developments.
- Sensors can provide powerful information about the health status and production efficiency in a poultry house.
- The use of robots may not be far away in the future.
- Integral data collection, clever data storage and early warning systems provide information to the farmer to improve results at all levels.
- Because the poultry chain exists of different sectors, data sharing between sectors is of importance to improve technical and financial results.
- The future challenge is to develop and support an open data infrastructure where sensor data and technical results are shared between links in the chain without claims and distrust.

5.1 Modern poultry chains

Poultry farming is mainly focused on either meat- or egg production and to a lesser extent to breeding and reproduction. The structure of the meat- and egg sectors are quite similar as depicted below (Figure 5.1). However, the management and techniques to obtain good results may differ largely between both sectors. Each sector and each level has its specific challenges to maintain production, welfare, animal health and economic returns at the highest possible level. Techniques to support decision making in the management of birds may differ between sectors and levels but can be of high value when used at the right place and moment. The similarities and differences between the two sectors are described below.

5.1.1 The broiler meat production chain

The 'primary breeding sector' consists of only a few companies that breed pedigree stock. Pedigree stock ('pure line') is kept on high-level bio secure farms. Eggs are hatched in a special pedigree hatchery and their progeny goes to the great grandparent (GGP) and grandparent (GP) generations. The eggs of the GP go to a special GP hatchery to produce Parent Stock (PS) which passes to the production sector. A single pedigree-level hen might have 25,000 parent stock bird descendants, which in turn might produce 3 million broilers. As of 2017, only two main breeding groups remained: Aviagen (with the Ross, Hubbard, Arbor Acres, Indian River and Peterson brands) and Cobb-Vantress (with the Cobb, Avian, and Sasso brands).

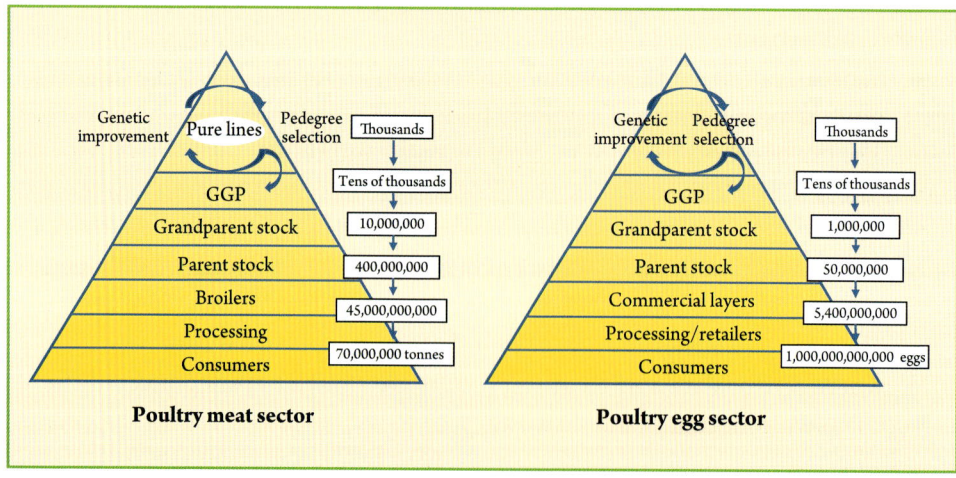

Figure 5.1. Overview of the poultry meat- and egg sector (after Mackay, 2008).

Broiler breeder farms raise parent stock which produces fertilised eggs. The males and females are separate genetic lines or breeds so that each line can be selected for optimal traits for productivity in either females or males, rather than a single line in which a compromise is reached between female and male optima. The chicks they produce will therefore be crossbreeds or 'crosses'. Since the birds are bred mainly for efficient meat production, producing eggs can be a challenge (Decuypere *et al.*, 2001).

Broiler breeder business is typically a two-stage process. Parent stock purchased from a primary breeder is delivered as day-old chicks. Most are placed in specialist rearing houses first until approximately 18-20 weeks of age, when they will be moved to a laying farm. The average broiler breeder farm houses 15,000-30,000 birds that start laying hatching eggs around 23 weeks of age. Each female bird will produce about 150-180 hatching eggs with an average hatchability between 80 and 85% resulting in roughly 120 to 155 chicks. They produce for a period of 34 to 36 weeks until the birds reach around 60-65 weeks of age and are slaughtered for meat. This cycle is then repeated when another flock of 20 week-old birds is placed into the house to begin the process again. As a general rule, each breeder farmer produces enough broiler hatching eggs to supply chicks for 8-10 broiler farms. Generally, breeder flocks are either owned by family farms (N-W Europe), integrated broiler companies or hatcheries or are contracted to them on a long-term basis.

Hatcheries receive the fertilised eggs, incubate them, and produce day-old broiler chicks. Incubation takes about 21.5 days and is often a two-step process. Initial incubation is done in machines known as setters. Eggs are held relatively tightly (large end up) in trays, which are stored on trolleys, typically containing around 4,800 eggs. One commercial setter can hold 60,000 to over 100,000 eggs. Inside the setter, temperature and humidity are closely maintained to keep the eggshell temperature close within the range of 37.6-38.2 °C and to facilitate 10-12% egg weight loss during the first 18 days of incubation. Blowers or fans circulate air to ensure uniform temperature, and heating or cooling is applied as needed by automated controls. The eggs are turned automatically on an hourly basis. The setter phase lasts about 18 days. Then, the eggs are removed from the setters and transferred to hatchers. These machines are similar to setters but without the turning device. When eggs are transferred from the setter to the hatcher, they are often candled first to remove clear infertile eggs and eggs with dead embryos. In the hatcher, the eggs are placed in baskets so they can rest on their sides, and newly hatched chicks can walk in the basket. Having a separate machine helps to keep hatching debris out of the setter. The environmental conditions in the hatcher are optimised to help the chicks hatch. As a commercial example, a large hatcher has a capacity of 15,000-40,000 eggs.

Some incubators are single-stage, and entire trolleys of eggs can be rolled in at once and are taken out after 18 days of incubation. One advantage of single-stage machines is that they can be thoroughly cleaned after each setting or hatch, whereas a multi-stage setter is rarely shut down for cleaning. In multi-stage incubators, eggs of different ages are placed together where the fresh eggs cool down the eggs with heat-producing embryos to avoid overheating. Warm eggs with older embryos heat the fresh eggs. The single-stage environment can be adjusted for eggs from different producing flocks, and for each step of the hatch cycle. On hatch day (day 21), the trays are removed ('pulled') from the hatchers, and chicks are removed from the trays. Chicks are inspected, with poor quality chicks being disposed of. Chicks may be vaccinated, sorted by sex, counted, and placed in chick boxes. Stacks of chick boxes are loaded on trolleys and placed into trucks for transport and arrive at the broiler farm within a few hours. Specialised climate-controlled trucks are typically used, depending on climate and transport distance.

At broiler farms, the chicks grow to slaughter weight in between 30-42 days (regular fast-growing broilers) or 56-84 days (slow-growing broilers). Broiler houses can be as wide as 15-20 m and over 100-120 m long. Day-old chicks from the hatchery are placed on pre-warmed floors with sufficient bedding materials in the entire house or brooding rings. All broilers have unrestricted access to feed and water, the climate is well-controlled according to their requirements. Different welfare concepts exist, in where chicks are housed at different stocking densities ranging from 10-20 birds per m^2. Houses with extra outdoor facilities and other rest and activity enrichments may exist for the non-traditional slower-growing broiler concepts. When the birds are full-grown, they are caught manually or automatically with harvesters and placed in crates or containers and transported by truck to a processing plant.

Broiler production is either integrated or non-integrated. In the first case, a single company (the integrator) controls several or all links in the production chain. Many integrators contract the growing process to individual growers, who are paid a set fee for the facilities, labour and non-feed variable costs. Day-old chicks and feed are delivered by the integrator, and slaughter ready birds are collected to be slaughtered in the slaughterhouse. The integrated model is found in Austria, France, Germany, Italy, Spain and the UK in particular. In non-integrated production, the different links in the chain are independent companies trading on the market. This means that breeders and growers buy feed and birds at their own risk, and are directly exposed to market fluctuations. Production tends to be non-integrated in Belgium, Finland, the Netherlands, Poland and Sweden.

In 2017, the total poultry meat production in the EU-28 was 15.9 million tons (MEG, 2018). The main poultry meat was broiler meat (81% of the total poultry meat production), followed by turkey meat and duck meat. The total number of farms with poultry in the EU-28 is more than 2 million. Of these farms, only 19,000 farms can be described as professional because they have more than 5,000 birds. In the poultry meat supply chain, different companies are involved in supplies (e.g. hatcheries, feed mills) and slaughter/processing. According to Van Horne (2018), the supply chain needs approximately 23 full-time workers for the production of 1000 tons of poultry meat. The estimated total employment in the EU poultry meat industry is 367,000 employees. The total production value of the poultry meat sector in the EU-28 in 2017 was almost €38 billion. This is the total value of the production at the primary farms, the slaughterhouses and the further processing of the poultry meat. The EU is an important player in the international trade of poultry meat. In 2017, the EU-28 exported 1.7 million tons of poultry meat with a value of €2 billion. At the same time, the EU-28 imported 0.8 million tons with a value of €2 billion (Van Horne, 2018). A relatively high volume of low-value products is exported, as legs, etc., whereas a lower volume of higher value products is imported, as filet meat from countries with lower production costs.

5.1.2 The layer egg production chain

The production chain for table eggs or industry eggs differs from the production chain of broiler meat (Figures 5.1 and 5.2). The more than 400 million laying hens in the EU produce over 7.5 million tons of eggs annually, three quarters of which originate from seven Member States: France, Germany, Italy, Spain, the United Kingdom, the Netherlands and Poland (Van Horne, 2019). About 10% of production consists of hatching eggs. In 2017, the total egg production for consumption in the EU-28 was 6.8 million tons. The EU is the world's second-largest producer of eggs after China and a net exporter of eggs and egg products. It has a self-sufficiency of around 105%. The seven leading countries each produce more than 500,000 tons of eggs. The total number of farms with laying hens in the EU is 3.9 million. Of these farms, 11,740 farms have more than 3,000 hens and can be described as commercial.

Figure 5.2. The table egg layer chain and the broiler meat production chain. Photo courtesy of Hendrix Genetics, Royal Pas Reform, HAS University of Applied Sciences and AERES University of Applied Sciences.

In the egg supply chain, different companies are involved in supplies (e.g hatcheries, feed mills), packing and processing. The total employment in the EU egg supply chain is estimated to be 186,000 full-time workers. The total production value of the egg sector in the EU-28 in 2017 was €8.2 billion, at farm-level prices. The total production value at retail prices is €14.5 billion. The EU is an important player in the international trade of eggs and egg products. In 2017, the EU-28 exported eggs and egg products with a value of €221 million. At the same time, the EU-28 imported eggs and egg products with a value of €31 million (Van Horne, 2019).

There are four main systems for keeping laying hens in the EU. In 2018, half of the EU's hens were kept in enriched cages – cages equipped with perches, nests, scratching areas and nail shorteners, which replaced the conventional battery cages banned by the EU in 2012. The other half were kept in cage-free egg production systems: either barn, free-range or organic systems. Barn systems are large enclosures with litter on the floor and freedom of movement for the birds within the poultry house. Each concept has its specific requirements and regulations for stocking density, total numbers per house, housing and management. Free-range systems are similar to barn systems with access to an outdoor run with a minimum of 4 m² per bird. The UK, Ireland and Austria have the highest shares of laying hens kept in free-range systems, while Denmark and Sweden are the countries with the highest shares of hens kept according to organic standards.

The organisational structure in the egg sector is very different from the poultry meat sector and varies greatly between countries. There is a link between housing systems for hens, farm size and level of production chain integration (either semi-integrated or with no coordination). The egg production chain is less integrated than the broiler meat production chain. However, in many EU Member States, such as the UK, Germany, Spain, Italy and Poland, egg production is widely integrated, with large companies sometimes keeping more than a million laying hens in cage systems. Substantial portions of the chain are integrated, which means that pullet rearing, layer management, feed supply, packing, processing and marketing to the retailer are carried out by a single company or cooperative. Feed mills often play an important role in egg production, providing producers with feed, pullets and advice. This is the case in France and the Netherlands. In most EU countries, there are independent layer farmers who bear the full risk of changes in input (feed and pullet) and output (eggs) prices. In countries, such as Greece, Portugal and Romania, the egg production structure is non-integrated and fragmented. In the egg production chain, the various activities – breeding, hatching, rearing and egg-laying – often take place on different farms to prevent the possible spread of diseases. Layer farms, especially the large ones, often include the egg grading and packing activities preceding delivery to the retail or wholesale market, to food services (restaurants, catering companies and institutions) and the food industry.

5.2 Techniques to monitor welfare and performance

With the increased intensification of the broiler and layer industries, with progressively fewer farms but with increased numbers of animals per farm, and the tendency towards increased awareness for animal welfare, the urge for automated systems and optimal control increased as well. Therefore the need for sensor and monitoring systems increased at the end of the 20th century, and this kick-started a new area of research and development of applications. Since the introduction of Precision Livestock Farming (PLF) at the start of the 21st century (Berckmans, 2017; Werner et al., 2003), the technologies of PLF aim to manage the growth of individual animals to create added value by real-time supervision of environmental factors, animal health and yield, production, reproduction, and welfare in an automatic, continuous, and non-invasive form throughout the production process, without any additional stress on the animals. Researchers have devised warning systems to alert farmers about immediate actions to deal with unexpected situations (Wathes *et al.*, 2008). Some expert knowledge has been computerised to offer a consistent, inexpensive, fast, and steady decision making support tool based on reproducible and accessible knowledge without any external influences (Mollo *et al.*, 2009). The technologies of PLF have focused on not only the factors that influence the production process but also on the most significant participants in the production process: the animal (Berckmans, 2017).

The emergence of PLF stimulated the overall expansion of poultry production, within which chickens were tracked per flock. The application of advanced technology is commonly seen in the poultry houses, such as animal inspection with sensing technologies like image and sound analysis, and the assessment of the relationship between the behaviours, welfare of chickens and environmental impacts with preference test (Norton *et al.*, 2016). Next to techniques to monitor poultry welfare, PLF can also be applied to monitor and improve technical results and production efficiency.

5.2.1 Poultry welfare

An overview of the current state of PLF research in poultry farming is given in a review by Rowe *et al.* (2019). The interest in PLF technology for poultry farming did not take off until the late 2000s, reaching a peak during the years 2017-2018, suggesting increased interest and investment in poultry PLF research and outcomes. Reviews on PLF technologies in poultry farming as by Mollo *et al.* (2009), Corkerey *et al.* (2013), Ben Sassi *et al.* (2016) and Astill *et al.* (2018) focused on technological developments in this field and ignored bird welfare developments. Evidencing the continued interest and investment in PLF technologies in the poultry sector, the Foundation for Food and Agriculture Research (FFAR) and the McDonald's Corporation has recently launched 'SMART Broiler', a research grant of $4 million to drive the development and commercialisation of automated monitoring tools to assess broiler welfare (Johnson, 2019). To use behaviour as a welfare measure, the behaviours must be validated to show whether and how they are linked to an animal's welfare status. PLF technology can be used to monitor many parameters, such as behaviour, but whether and how the measurements are linked to a parameter, i.e. the internal validity of the measure is not always clear. Furthermore, whether and how the monitored parameters are linked to welfare, i.e. the external validity of the measure, must be further established too (Winckler, 2019). Fernandez et al. (2018) compared locomotory behaviour data to other measures collected via a validated assessment protocol (Welfare Quality®). They found statistically significant correlations between locomotory behaviour (activity and occupation patterns) and welfare scores (for footpad lesions and hock burns), indicating that activity and occupation patterns are valid indicators of broiler welfare status. While there is the potential for PLF to improve standards of bird welfare through continuous, real-time monitoring, important steps as internal and external validation still need to be taken (Rowe et al., 2019; Van Erp-van der Kooij and Rutter, 2020).

Poultry farming, and in particular broiler farming, is an important area to make efforts on improving welfare, because of the large number of animals involved and the potential for improvements in their welfare (Bennet et al., 2018). Broilers are the world's most numerous bird, with a standing population of 22.7 billion, an order of magnitude greater than the standing stocks of any other farmed species. The total impact of small welfare improvements for large numbers of animals may be larger than the impact of large welfare improvements for few animals. More than a decade ago, modern broilers suffered from problems such as sudden death syndrome, ascites, lameness and contact dermatitis as a result of their fast growth rate (Bessei, 2016; Knowles et al., 2008), which has increased through breeding programs focused on increased breast meat yield by 400% since the 1960s (Zuidhof et al., 2014).

Broiler breeder farmers face challenges to maintain both high hatching egg production rates and welfare levels (Decuypere, 2001; De Jong and Guémené, 2011). Where the motivation to invest in PLF systems in the final products (broilers and layers) may be low, breeding organisations make more use of PLF systems to support their highly valued breeding stocks. Their knowledge and expertise with PLF may be of use to introduce these techniques in broiler and layer farms on a larger scale as well. Numerous techniques are used to assess the pedigree stock. For example, birds might be examined with ultrasound or x-rays to study the shape of muscles and bones. The blood oxygen level is measured to determine cardiovascular health and the walking ability of pedigree birds can be observed and scored (Stevenson, 2017). On the lower end of the production pyramid, however, each individual animal is worth relatively little and the turnover of flocks is very fast, with modern broiler strains in conventional intensive production systems reaching their target weight in just 5-6 weeks or sometimes even less (Wilhelmsson et al., 2019). This means that concern for the welfare of an individual bird may be low. The profit margin for poultry farmers having final products is small, creating further conflict between production, bird welfare, health and environmental issues (Eijsakkers and Scholten, 2011).

While the number of chickens farmed for meat is already huge, global meat consumption is currently predicted to increase, not only because of a growing human population but also because of increasing incomes and urbanisation (Scholten *et al.*, 2013). This means that poultry farming is expected to increase in low-income countries where animal welfare may not yet be seen as a priority. Poultry meat consumption specifically has grown in comparison to other meat types. Where beef, pork and sheep meat consumption levels have varied very little between 1990 and 2017, poultry meat consumption has increased by 70% (OECD Meat Consumption, 2019). Therefore, poultry welfare is an important area to focus on and to make efforts in improving welfare, and PLF is one potential tool to achieve this (in addition to improving farming practices in other ways), through enabling continuous monitoring and fast interventions benefiting individuals in their lifetime. PLF likely has the potential to improve both bird welfare, health and production efficiency in the poultry sector. The introduction of PLF techniques to improve animal welfare would be boosted when technical and hence financial results would be increased as well.

Widespread public concern and debate about the welfare of laying hens resulted in a ban on conventional cages in the European Union on January 1st, 2012, ultimately resulting in a substantial increase in cage-free egg farms. More recently, many multinational hospitality companies, restaurant chains, and major retail concerns have made pledges to the use of cage-free eggs only. This has led to an increased demand for cage-free egg production worldwide. Cage-free housing systems may offer a better laying hen welfare compared with caged systems, mainly because of increased useable space and opportunities to perform highly-motivated behaviours (Appleby and Hughes, 1991; Duncan, 2001; Freire *et al.*, 2013; Rodenburg *et al.*, 2012; Shimmura *et al.*, 2010). On the other hand, the welfare of hens in commercial cage-free systems may be compromised due to high flock mortality, feather pecking and cannibalism, and keel bone and footpad disorders. The physiological capabilities of laying hens, the required abilities to move and perform in cage-free housing systems, management procedures, housing design and feeding strategies may also lead to some of the above-mentioned welfare problems (Heerkens *et al.*, 2015, 2016a,b). The gradual transition to cage-free systems on a global scale emphasises the importance of PLF as a potential tool to utilise the potential welfare benefits of cage-free housing systems while safeguarding production efficiency.

5.2.2 Production performance

Group numbers and uniformity is a challenging subject in poultry farming. Where the focus on dairy farms and pig farms is more on the individual animal or small groups, commercially housed chickens are always placed in large numbers in flocks of 10,000-80,000 birds per unit and flock uniformity is of the highest importance to gain optimal technical results. Therefore not only a vast understanding of the requirements of individual chickens is needed, but also flock dynamics and uniformity measures in both animals and environmental conditions are very important (Lokhorst and Ipema, 2010). Next to uniformity, also economic aspects as stocking density and feed prices will play an important role in decisions to optimise technical results and to maintain high levels of animal welfare at the same time. It was only in recent years that in Western Europe the abusive use of antibiotics is banned due to increased resistance of bacteria (MRSA, ESBL) to treatments of bacterial infections in human. Further, due to increased awareness for animal welfare, other types of housing systems emerged that may challenge bird health, such as aviary and outdoor or free-range systems instead of cages.

To produce high numbers of healthy broilers and layers, the structure and capacity of the poultry sector are changing at an equal pace with increased capacity and numbers of hatcheries and breeder farms (Figure 5.1). However, the number of primary breeder organisations (pedigree and (great-)grandparents) supplying the commercial hybrid parent stocks, decreased and intensified their breeding activities. In integrated poultry chains, the reward to introduce upstream measures or techniques to improve the overall profits at all subsequent levels (breeder, hatchery, production farms, slaughterhouse) will be large. Upstream investments may pay out downstream in the broiler or layer sectors, but in transparent healthy poultry chains, the profits will return to the upstream companies as well. In non-integrated chains, the incentives to improve egg- or chick quality and reduce antibiotics use will be lower and hence the chain failure costs in non-transparent chains can be immense (Lourens and Steentjes, 2008; Lourens et al., 2011). This points to the challenging possibilities of combining PLF techniques at individual farms and a functioning integrated database to monitor the technical and financial results of the whole chain.

5.3 Application of sensor techniques in poultry farming

5.3.1 Use your senses

PLF may prove to be of value in poultry farming, but a thorough understanding of the basics of bird behaviour and climate cannot be overestimated. With all techniques available to monitor all kinds of data inside and outside the farm, it is always advised to make important management decisions based on the biology and responses of birds by using your own sensors first: vision, smell, sound, feeling, taste, etc. Spend time on the farm and use your senses and gut feeling. This principle of 'look, think, act' is the backbone of the informative and well-illustrated books about poultry farming developed by Roodbont (http://www.roodbont.nl) and co-workers.

5.3.2 Environmental control

Building materials, equipment, environmental control and management systems (stocking density, curtains, equipment, and nutrition) influence the microenvironment around individual birds. The spatial distribution of these variables may indicate stress zones within the house. Miragliotta *et al.* (2006), studied the spatial distribution of environmental temperature, air, and noise conditions inside a broiler house with tunnel ventilation for a flock with a stocking density of 18 birds per m^2. It appeared that the stress zones were located at both ends of the house, and the highest mortality rate was found in the west, where the exhausting system was located. Spatial analysis may be an important tool to determine the performance of ventilation and cooling systems, as well as air quality inside broiler houses. Using a dynamic flow chamber to measure ammonia emissions from poultry litter and to investigate ammonia emissions responses, Liu *et al.* (2007) observed that ammonia emissions were very sensitive to litter moisture. Total nitrogen content in the air increased with increasing litter humidity, indicating that this may potentially increase ammonia emissions. However, measurements of ammonia levels and total nitrogen losses in litter suggested that when water was applied, ammonia emissions were suppressed for a short period. These results may allow researchers to develop joint systems for ammonia emission analysis, enabling mitigation measures to be taken by the algorithms implemented in the automatic controllers when levels become critical.

5.3.3 Sensors to measure animal-based responses

Until 20 years ago, the hatchery business changed from controlling climate conditions towards the control of the direct environment around the embryo: the egg. It was realised that air temperature is not the same as egg or embryo temperature (Lourens, 2001). To support modern hatcheries with advice and equipment to hatch healthy chicks, incubator manufacturers developed many devices that measure embryo responses concerning the environment.

Based on the incubation experiments by Lourens *et al.* (2005) who showed the importance to measure eggshell temperatures instead of machine air temperatures in incubators, the company Petersime developed the OvoScan™ system (Figure 5.3). Using this device, the incubator temperature is continuously adjusted in response to the actual eggshell temperature. The responses of embryos to external conditions was taken further in both scientific as applied incubation research. Bamelis *et al.* (2005) introduced the concept of acoustic monitoring of the hatching process. Similarly, Exadaktylos *et al.* (2011) used embryo vocalisation to accurately identify the time at which 93 to 98% of the embryos penetrated the internal air cell (internal piping). The time difference between the hatch times of the first and the last hatched chicks is called the hatch window (HW). Practically, accurate HW measurement is difficult because this requires the monitoring of hatch times of individual chicks. Decuypere *et al.* (2001) reported variability in HW of 24 to 48 hours, mainly depending on variation in embryo temperature and variation in egg characteristics as breed, breeder age, egg size and egg storage time (Lourens *et al.*, 2007). The main issue is that early hatched chicks often have no access to feed or water up to 72 hours after hatch by considering the spread in HW, chick handling, and transport time, which is critical from the welfare aspect for both day-old chick quality and post-hatch performance (Hulet, 2007). To detect and follow the hatch curve, Petersime (Belgium) developed the Synchro-Hatch™ device (Figure 5.3) and used embryo-response technology to synchronise hatch profiles to the hatching process (Romanini *et al.*, 2013). It automatically detects the exact timing of 100% internal pipping (IP) and then initiates a sequence of modifications to the incubation environment to stimulate simultaneous hatching. The technique also automatically recognises when all chicks are hatched. The system triggers a further phase in the incubation process to optimise the finishing of the chicks to prepare them for take-off. These actions reduce hatch time and concentrate the hatch much nearer to the time of chick take-off.

Figure 5.3. The OvoScan™, Synchro-Hatch™, the CO$_2$NTROL™ and the Dynamic Weight Loss System™ by Petersime (Belgium) to support the embryos during the incubation process (photo courtesy Petersime N.V.).

Gaseous exchange is known to be one of the most essential incubation factors. To drive the metabolism towards developing a healthy chick, oxygen has to be supplied to the egg and carbon dioxide has to be removed as a waste product. Consequently, maintaining the correct levels of CO_2 during the entire incubation cycle has a beneficial effect on the development of the cardiovascular system of the embryo. This level can be controlled by adapting the ventilation. Alongside enhancing embryonic development in setters, precisely timed CO_2 stimulation in hatchers leads to simultaneous pipping and hatching as well as improved chick quality. CO$_2$NTROL™ takes care of the online measurement of CO_2 levels as input for ventilation control (Figure 5.3).

The fourth important feature developed by Petersime is the Dynamic Weight Loss System™ (DWLS™) that adjusts the humidity level in the incubator based on measurements of weight loss of the eggs during the incubation process. The placement of the DWLS™ within a trolley of eggs is illustrated in Figure 5.3. Eggs need to lose a certain amount of weight during incubation to achieve optimal hatchability and day-old chick quality. This requires water to be transported from the egg to the environment via the eggshell. By controlling the humidity level in the incubator, the rate of this water (and weight) loss can be managed, taking into account the eggshell water vapour conductance of the specific batch of eggs.

It is easier to monitor live chickens than embryos in eggs. According to Rowe *et al.* (2019), in most poultry PLF studies, cameras or sensors such as climate sensors and wearable sensors are used. This included RFID; although RFID is used for individual identification of animals, this technology can also be used as movement sensors (Ben Sassi *et al.*, 2016), and used to track behaviour, including locomotory behaviour. Microphones seem less popular (Neethirajan, 2017).

5.3.4 Weighing scales

Whatever data is collected to monitor performance, health status, activity patterns, behaviour or welfare and climate, the monitoring of body weight and growth should be the basis to start from. Individual body weights are very important for a flock manager. Weight, and especially weight changes provide insight into the physical condition of the flock and its future performance. It can be used to slow down growth in fast-growing broilers allowing them to develop a better heart, blood system and strong bones (Roush and Wideman, 2000). Weight and uniformity are also important to determine the slaughter moment and slaughter yield. Furthermore, to achieve optimum performance, pullets and breeders must grow to specific bodyweight standards, varying between breed types and seasons. Generally, birds should be weighed weekly starting from the first week. During the first 2 weeks, birds can be weighed in groups or batches in buckets. After this, birds should be weighed individually. As a rule of thumb, 2-5% of a flock and never less than 75 birds should be weighed on the same day of the week, at the same time. Weighing a sufficient amount of birds, and all birds penned, is the only way of assessing uniformity. According to Ken Laughlin (pers. comm.), you have to weigh birds until the addition of one extra bird does not change the cumulative average.

During the nineties of the last century, Alwyn Havard from NEDAP developed the Peckode® system to measure individual body weights of male and female breeders separately. This allowed gender-based weight management and controlled feeding which improved technical results enormously. A separate male and female weighing system by using the Peckode® leg band identification device on the male bird was developed (Figure 5.4).

Unfortunately, this system is not commercially available any more. The investment in automatic weighing systems though seems to pay back quickly, allowing real-time monitoring and early warning signals to avoid (re)production problems. Automatic scales save labour, especially when using more than one platform, and offer the key

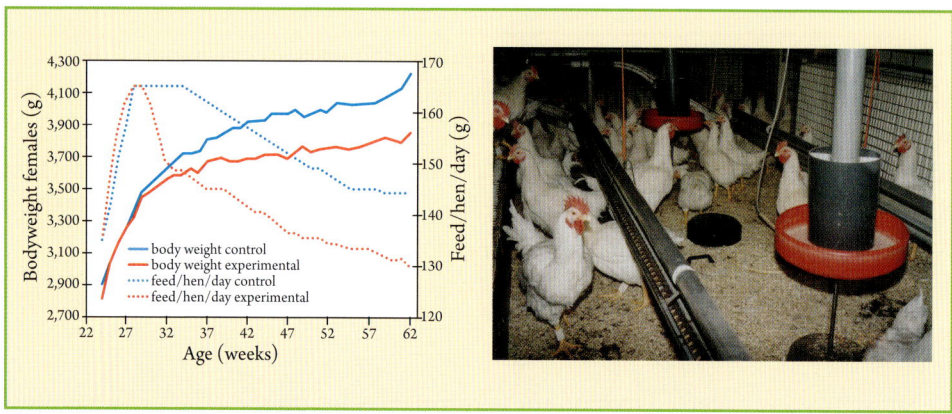

Figure 5.4. Experiments with broiler breeders using the Peckcode® System at research institute 'Het Spelderholt' (the Netherlands) (photo courtesy Nedap).

to light and feed control. Experience in the field showed that bird age and behaviour determine the accuracy of the weightings, often leading to an underestimate of body weight for older and less active birds.

For poultry, the nutrient composition of the diet is often based on the feed intake and nutrient requirement of the bird with the highest requirement, including some safety margins. In this way, the total flock can be fed. However, this also suggests that some birds, probably most of the flock, will eat more nutrients than they require. The nutrients that are not used by the birds will be excreted, contributing to pollution and affecting the environment. It is better to feed according to the requirement, e.g. to apply precision feeding: meeting the nutrient requirement of animals as accurately as possible in the interest of a safe, high quality and efficient production while ensuring the lowest possible load on the environment (Banhazi et al., 2012a,b).

Even though a flock of chickens is considered genetically identical, there is still a substantial difference in body weight among individual pullets at the end of the rearing phase, which already suggests that nutrient requirements will also be different. This is supported by several recent studies that showed large variations in feed intake among laying hens within one flock. An innovative research tool (and in the future a potential commercial tool) is the Precision Feeding System, being developed at the University of Alberta by the group of Dr. Zuidhof (Figure 5.5). The PF system is a feeding station for individual birds where feed intake and body weight is monitored. Moreover, feed intake (according to the bodyweight curve) of individual chickens can be adjusted and controlled. With this system, uniform flocks can be produced with

5. Sensors and techniques to monitor and improve welfare and performance in poultry chains

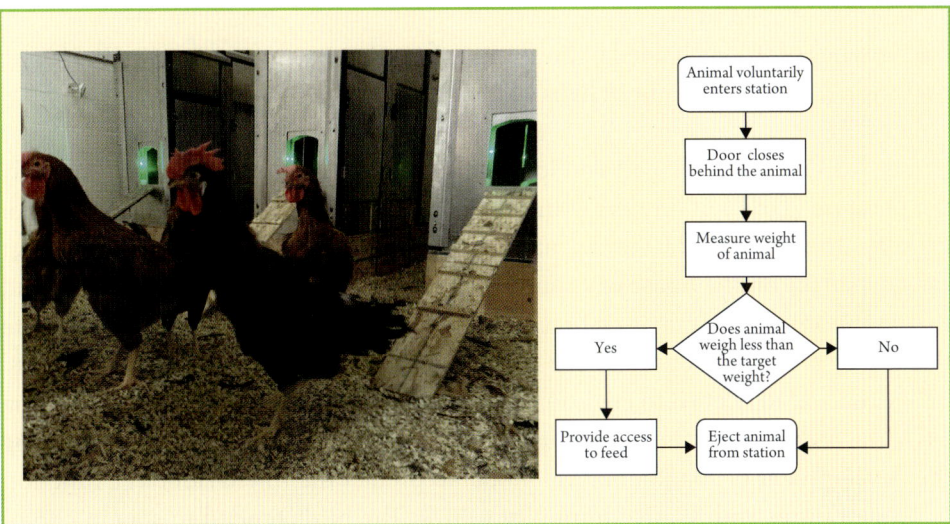

Figure 5.5. Precision feeding system developed at the University of Alberta (Zuidhof et al., 2017).

coefficients of variation as low as 3% body weight (Zuidhof et al., 2017). However, not all reared breeder birds came into egg production during adult life, indicating that there is variation in growth and development within a flock and that there is almost certainly a different nutritional requirement among birds (Girard et al., 2017a,b; Zuidhof, 2018).

5.3.5 Electronic identification

In most animal production systems, animal identification is essential. New technologies or those commonly applied in other industries can be used for monitoring and managing production. In modern broiler production, electronic identification (EID) technologies have been studied especially in behavioural studies under different housing conditions. Pereira (2003) identified the limits of the thermoneutral zone of broiler breeders using EID. This tool allowed the identification of birds that were more resistant to the harsh weather (essentially hot and dry). Despite being effective to quantify behaviour, EID did not allow the classification of the recorded behaviours. For example, the time difference between an RFID-tagged bird passing two RFID readers and the distance between these readers enables movement speed to be calculated, and behaviours such as time spent feeding and resting can also be monitored (Zhang et al., 2016). As another example, RFID has been used to sense when a hen enters or exits a nest box which, along with a pressure sensor to detect the presence of an egg, has led to the design of a smart nest box to monitor the laying performance and behaviour of hens (Chien and Chen, 2018).

Hyun *et al.* (2007) developed a radio telemetry system devise to measure deep body temperature (DBT) of broilers with a 0.1 °C accuracy, to predict the occurrence of diseases. The collected data of fluctuation of DBT were used to develop a system that can control the broiler house environment in real-time. Curto *et al.* (2002) suggested that monitoring and evaluating breeders' behaviour can be a useful tool to evaluate the quality of the environment. The objective of their study was to predict breeders' behaviour and to relate it to environmental temperature and relative humidity using EID. The path used by the female breeders inside the house was recorded, allowing the development of a mathematical model to predict their behaviour as a function of temperatures and relative humidity values. The birds tended to stay in cooler places when temperatures increased (Bottje *et al.*, 1983; Yanagi *et al.* 2002). There is an increasing need for real-time monitoring of broilers in commercial houses, given the complexity of the environmental variables that may affect their welfare. The use of specific systems to acquire and process images in real-time enables researchers to study bird behaviours in a non-intrusive manner and to observe and assess specific events with no human interference.

5.3.6 Early warning systems

The use of real-time data collecting systems was studied by Kristensen *et al.* (2006) who developed a dynamic model that could predict the activity of broiler chickens in response to changes in light intensity to control their activity in the future. In that study, they combined dynamic modelling techniques with traditional statistics to understand the underlying mechanisms that control the activity of broiler chickens. These techniques allow real-time monitoring of birds' activities, reporting them, and making changes in the poultry house equipment, including feeders, fans, and sprinklers, based on the recorded information. This real-time information can be used as Early Warning signals in case things go wrong. An example of how Early Warning signals can be used in broiler farms is the EyeNamic Activity Sensor technique developed by Fancom (Figure 5.6). It demonstrates unexpected or planned effects as light changes, vaccination moments, feed and water problems, and the moment of thinning.

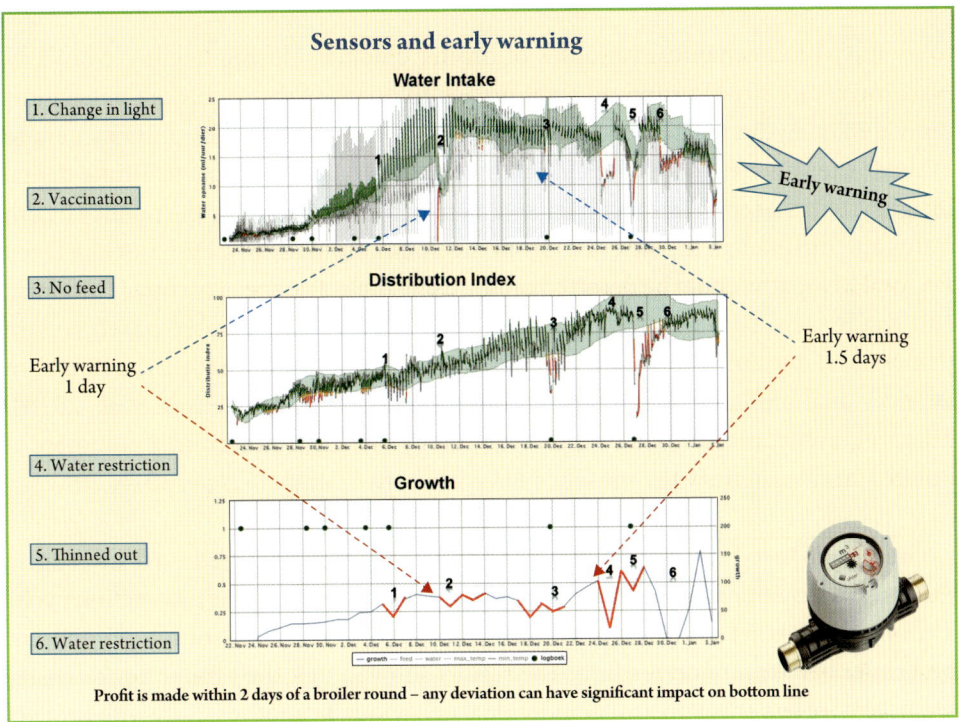

Figure 5.6. Example of water intake and eYeNamic distribution sensor output (by Erik Vranken; courtesy Fancom).

Most papers described prototype systems, suggesting that relatively few PLF systems are currently commercially available for poultry farms (Rowe et al., 2019). The commercially available technologies were: the eYeNamic™ camera system (De Montis et al., 2013; Norton et al., 2016), and environmental sensors to measure temperature (Carvalho et al., 2013; Coelho et al., 2016; Curi et al., 2017; D'Alfonso et al., 1996), ambient dust (Zhao et al., 2009), relative humidity (Carvalho et al., 2013; Coelho et al., 2016; Curi et al., 2017); vibration (Chen et al., 2010), ammonia concentration (Ji et al., 2019), carbon dioxide concentrations (D'Alfonso et al., 1996; Ji et al., 2019), and a thickness and crack sensor for eggs (Li et al., 2011).

5.3.7 Camera's and images

In recent years, animal welfare has gained attention among researchers. Animal welfare is commonly assessed based on mortality, physiology, behaviour, and health (Van Hertem et al., 2018). Animal behavioural parameters can be monitored with photo and video cameras to estimate welfare status in commercial broiler houses (Dawkins et al., 2013; Pereira et al., 2013). Mehdizadeh et al. (2015) used machine vision to identify biomechanical variables of broiler chickens during feeding. Similarly, Nakarmi et al. (2014) proposed an approach to quantify some chicken behaviours individually. Aydin (2017a,b) used images of broilers recorded by a 3D camera to evaluate the lameness of broilers. Zhuang et al. (2018) developed a real-time digital image system that could inspect the movement of broilers, monitor the poultry health status and distinguish the sick ones from the flock automatically. Diseases like Deep Pectoral Myopathy could be detected online using dielectric spectroscopy (Dominiak and Kristensen, 2017; Traffano-Schiffo et al., 2018). A 3D camera-based system that could determine the weight of several broilers at once or predict the weight of an individual broiler was developed (Amraei et al., 2017; Mortensen et al., 2016). Thermographic images based on IR technology were introduced in a commercial poultry farm to identify the presence of laying hens (Zaninelli et al., 2016, 2017).

Collins (2008) used video-capture systems to develop a tracking method to investigate the social behaviour and preferences of broiler chickens during the production cycle. The objective of that study was to assess the impact of density and social environment on broiler behaviour and feeding. The results showed that the broiler flock density seems to have less effect on broiler individual behaviour. In another study with broilers breeders, Pereira et al. (2012, 2013) used video cameras to perform a qualitative analysis of broiler breeder behaviour, to identify birds' behavioural responses to the environment that could indicate their welfare status. The results showed a direct influence of the rearing environment on the broiler breeders' behavioural responses. The presence of feed was also an experimental factor, and it was demonstrated that it influenced the performance of behaviours related to welfare. Dawkins et al. (2013) provided evidence that the optical flow patterns of broiler flock movements recorded on video or CCTV (Closed-circuit Television) inside commercial broiler houses can provide a potentially valuable estimation of broiler welfare. The authors proved that simple descriptive statistics of optical flow patterns (mean, variance, skewness, and kurtosis) differ between flocks and that these differences are correlated with more conventional assessments of welfare, such as mortality and percentage of birds with leg problems in a flock.

Nääs et al. (2009) proposed a methodology for estimating thermal comfort in broilers and behavioural responses related to welfare indicators were evaluated. An algorithm was developed for classifying specific behaviours of broiler breeders and then transferred to a software application. Videos were recorded within a climatic chamber, where temperature varied according to temperature limits and extremes within the range of thermal neutrality. The digital images were processed and allowed the detection of a sequence of welfare-related behaviours. The resulting images, after analysis, allowed visualising the clusters that presented determined behavioural patterns. This may be a step towards real-time recording and continuous welfare assessment of flocks.

5.3.8 Sound analysis

To assess animal welfare, vocalisation analysis is an interesting tool to collect data in a non-invasive manner. Moura *et al.* (2008) studied the relation between thermal environment and pullet performance. The idea was to estimate thermal comfort by analysing vocalisation magnitude and frequency of pullets reared in tropical conditions. The results showed a correlation between a pullet grouping pattern and vocalisation during thermal stress exposure. In the study of Woodcock *et al.* (2004) to determine if hen's calls at the feeder affect pullets' productivity and welfare, it was found that chicks exposed to recorded hen vocalisations were closer to the speaker than control chicks. This suggests that pullets try to interact with the hen they thought to hear. Hen vocalisation may improve growth during the first nine days post-hatch and efficiently mimicking natural stages may provide better rearing conditions for broilers and pullets in the starter phase.

Vocalisations of hens are an indicator of welfare status as it is affected by the corresponding environment (McGrath *et al.*, 2017). The division of Measure, Model & Manage Bioresponses (M3-BIORES) at KU Leuven (Belgium), developed a system to quantify the dynamic feed intake of broilers in different ages according to pecking sounds (Aydin *et al.*, 2015). They proposed a real-time pecking sound analysis tool to identify short-term group feeding behaviours (Aydin and Berckmans, 2016). Researchers at Georgia Tech Research Institute found that the number of vocalisations is a useful performance indicator for detecting stressful conditions. A support vector machine (SVM) model was proposed to separate chickens infected with infectious bronchitis (IB) from healthy chickens. Analysis of chicken vocalisation might be used to provide an early warning of unhealthy conditions for chickens. The features and algorithms could be adjusted to apply to different commercial settings (Carroll *et al.*, 2014; Rizwan *et al.*, 2016; Whitaker *et al.*, 2014). Similarly, Lee *et al.* (2015) presented an online monitoring prototype that could detect stress and classify it into types like physical and mental stress using vocalisation of laying hens. Microphone arrays of Kinects were used to automatically detect the anomalous status of laying hens through the number of vocalisations and area distributions during nights (Du *et al.*, 2018).

5.3.9 Vibration and pressure

Sensors to determine the impact of vibrations on the egg (micro) cracks from laying nests to egg collection systems as the Crackless Egg by Masitek (www.aaggrrii.com) are commercially available and widely used in many egg-producing systems. The accelerometer measures G-forces when eggs crash to each other or fall from one belt to the other. It provides a real-time measure of impact to the eggs, as well as the location and severity of the damage. One large impact or several small impacts may result in hairline cracks, responsible for high rejection rates or increased percentage of rotten eggs in the hatchery (Hein, 2007).

Pressure sensors have been used also to assess the walking ability of humans and animals. Nääs *et al.* (2008) conducted a research study that aimed at evaluating broiler locomotion ability. Birds were encouraged to walk on a platform, and the pressure of both feet was automatically recorded separately. The software application was set up in a desktop computer, and data were saved for further analysis, as used in earlier studies (Carvalho *et al.*, 2005; Oviedo-Rondón, 2007). Results showed that gait score increased as bird weight and age increased. Although not visually identified, the results using the measurement system showed that the peak force was different between the right and the left legs, causing slow and uneven walking, and further leg problems in older birds.

5.4 Future trends in precision poultry farming

5.4.1 Robotics

With the advancement of technologies and their potential applications in agricultural production, more intelligent machines, like robots, will be available to farmers in carrying out physical tasks in poultry production. New agricultural production concepts and expanded machine capabilities will open new scenarios of robotic applications, including precision livestock farming (Usher *et al.*, 2017). An in-depth understanding of 'system analysis and integration' will also be essential in providing holistic solutions to agricultural production systems, such as a whole poultry house management approach. There have already been many examples of the benefits that the 'system of systems' approach offers to engineering and agricultural systems (Ren *et al.*, 2020). As with many technological innovations, the advantages of robotics and machine capabilities enable us to substitute human labour and animal power in undesirable work conditions, as well as ensure uniformity and improve safety and productivity of the work performed. Some commercial examples include a robot for autonomously sanitising poultry houses, The Octopus Poultry Safe (OPS), ChickenBoy and robot chicken nannies for chicken health and welfare inspection, PoultryBot for picking floor eggs, and Spoutnic for training hens to lay eggs in nest boxes. These challenges will require the collaboration of scientists, engineers, and technologists from many disciplines. The current poultry production systems are very likely to benefit from innovations in robotics research and applications.

The Poultry Bot (Livestock Robotics, the Netherlands; Figure 5.7) could be used to assist farmers in critical daily tasks, such as detecting environmental conditions and collecting floor eggs (Vroegindeweij *et al.*, 2018a). The collected eggs could then be transferred to one side by rotating the end effector. Researchers introduced a path planning algorithm that could be evaluated through the information about floor eggs distribution probability. A result relevant to the tests is that the robot worked better in terms of the path structure and visiting times of high-potential locations (Vroegindeweij *et al.*, 2014). Monte Carlo Localisation, a localisation algorithm based on particle filters, was used to guide the PoultryBot tested in a simplified test site with the presence of live hens. Experimental results demonstrated a localisation accuracy of 0.37 m with an average error of 0.2 m (Vroegindeweij *et al.*, 2016) and a collection success rate of 46% using 300 test eggs (Vroegindeweij *et al.*, 2018b).

Figure 5.7. On the left PoultryBot, developed by the Farm Technology Group of Wageningen University, as a proof of concept of a mobile robot for applications in poultry houses in tasks like the collection of floor eggs. On the right, a prototype of CollEGGtor, as a continuation of PoultryBot's development by Livestock Robotics (www.livestockrobotics.nl). Courtesy of Livestockrobitics.

The Octopus Scarifier (Octopus Robots, France) is an autonomous machine designed to aerate poultry litter as well as monitor environmental measures such as shed temperature, humidity, ammonia levels and brightness (Figure 5.8). The company says that by scarifying litter, it can help to reduce diseases like aspergillosis, pododermatitis, hock burn and breast injuries. The device operates automatically and can run 24/7, with a charging unit located in the shed that it can return to when batteries are running low.

The Spoutnic (Tibot, France) is a smaller device than many others on the market and is designed to keep birds active in broiler breeder sheds, which encourages the use of nest boxes and may improve fertility. The company says its small size makes it manoeuvrable around poultry sheds and it can tackle uneven litter. It automatically avoids objects, weighs 10 kg and has an eight-hour battery. The machine has six-speed settings, from slow to fast which can be selected depending on the training phase of the flock.

The ChickenBoy (Faromatics, Spain; Figure 5.8) takes a slightly different approach to automation. It is suspended from the ceiling in poultry sheds and monitors shed conditions from above. The device uses cameras and other sensors to measure visual issues like dead birds or poor litter, can measure sound levels and other environmental variables like temperature, airspeed or CO_2.

Figure 5.8. The front view of the Octopus Poultry Safe (OPS) robot (left picture; source Octopus Robots) and the ChickenBoy robot at work (right picture; source FAROMATICS).

5.4.2 Precision feeding system

The Precision Feeding System as mentioned in Section 5.3.4 needs to be further developed with options to feed different diets. With this feeding system, individual feed intake can be determined, but also the birds preferences by giving them a choice of diets high or low in specific nutrients. Generally, birds are well capable to fulfil their nutrient requirement. This so-called choice-feeding will give insight into the specific requirements of birds related to their individual growth and development. This does not guarantee the best choice for growth, performance and welfare, therefore it is of interest to compare with diets differing in nutrient levels and a more practical diet. Daily feed intakes and growth of individual birds will be generated and can be used to study variation in nutrient requirement among birds. The high-resolution data collected with the PF system allows new insights into feeding behaviour and quantification of parameters of importance like estimates of metabolic rate and feed efficiency on individual birds in group-housing systems. Next to techniques to monitor and control individual feed intake, body weights and growth, emphasis needs to be paid to variation in developmental stages as well. The PF systems challenge us to look differently at poultry farming and move us to different forms of experimental set-ups with the individual bird as an experimental unit instead of pen experiments with group averages.

5.4.3 Data-driven based on a combination of sensor and management data

From the above, it may be clear that any PLF technique is most valuable and of practical relevance when it is introduced from a fairly good knowledge of poultry farming and a solid background in biological processes. Many prototypes were tested in scientific environments but disappeared before commercialisation in practice. We believe that most value is created from the combination of biological relevant sensors with fast and real-time data processing procedures. For that, more cross-overs need to be made between technical information sciences, applied animal sciences and environmental sciences. The application of PLF techniques at individual farms within a functioning integrated database to monitor technical and financial results of the whole chain will result in a reduction of failure costs and increased overall chain profits.

PLF makes use of technological resources to individually identify and process data, improving accuracy and allowing higher productivity (McBratney *et al.*, 2005). Equipment and systems need to be integrated to allow supporting each decision in particular. The information collected daily, along with systematic monitoring of production stages, produce important statistics for evaluating, controlling, and promoting possible improvements (Kebeler and Schiefer, 1996). Examples of software are the packages from FarmResult, OptiFarm or Sommen. PoultrySoft developed integrated Enterprise Resource Planning (ERP) software packages to register all transactions within the poultry chain between the different sectors and surrounding industries.

In Table 5.1, an overview of Smart Farming techniques for poultry is given concerning sensors, animal species concerned, where the sensors are located, what they measure and why, if the sensor techniques are available as commercial products and the company's website.

Table 5.1. Overview of smart farming techniques for poultry.

PLF sensor	Animal species	Where	What it measures	Why	Commercial product	Company website
Accelerometer	laying hens	leg tag	activity	locomotion	no	-
Sound analysis	laying hens	poultry unit	vocalisations, rale sounds, gakel calls	health, stress	no	-
Sound analysis	broilers	poultry unit	feed intake	production, health	no	-
Sound analysis	broilers	poultry unit	weight	production, health	no	-
Vision	broilers	camera	locomotion	lameness	no	-
Vision	broilers	camera	bodyweight	production, health	no	-
Vision	broilers	camera	behaviour	production, health	no	-
Vision	broilers	camera	posture	health, influenza	no	-
Vision	broilers	camera	distribution and activity	health, stress	yes	Fancom.com
Positioning	laying hens	rfid leg tags	locomotion, resting	health, activity	no	-
Positioning	laying hens	uwb sensor in a backpack	locomotion, behaviour	health	yes	Sensolus.com

>>>

Table 5.1. Continued.

PLF sensor	Animal species	Where	What it measures	Why	Commercial product	Company website
Weighing device	poultry farm	weight	growth, activity	production, health	yes	Bigdutchman.com; Fancom.com; Veit.cz; Opticon-agri.com; Sodalec.fr; Choretime.com
Precision Feeding Station	breeders, layers	uniformity	bodyweight	hatching egg production	no	xanantec.com
Ovoscan	embryos	eggs	embryo temperature	hatchability and chick quality	yes	petersime.com
Dynamic Weightloss	embryos	egg tray	egg weight loss	hatchability and chick quality	yes	petersime.com
Synchro-Hatch	embryos	hatcher basket	hatch curve	hatchability and chick quality	yes	petersime.com
CO_2NTROL	embryos	incubator	ventilation	hatchability and chick quality	yes	petersime.com
Software	broilers	poultry house	technical data	performance, management	yes	farmresult.com
Software	broilers	poultry house	technical data	performance, early warning, management	yes	optifarm.co.uk
Software	broilers, hatchery	broiler house, hatchery	climate, technical data	performance, management, climate control	yes	www.sommen.nl
Software	breeders, hatchery, broilers, slaughterhouse	poultry chain software	planning, production data	chain management, optimisation	yes	qfsoft.nl/ poultrysoft
Octopus Poultry Safe	layers	sanitising, disinfection	safety	hygiene	yes	octopusrobots.com
ChickenBoy	broilers	climate	management	climate	yes	poultry.proultry.com
Spoutnic	layers, breeders	floor eggs, dead birds, climate	object location, climate	labour efficient	yes	tibot.fr
PoultryBot	layers	floor eggs, climate	egg location, climate, temperature	labour efficient	yes	livestockrobotics.nl

The sensitivity of open data exchange between sectors in the poultry chain remains a challenge. Integrated companies within a poultry chain will not be reluctant to share data. This may be different between private companies within the chain that depend on each other. Different sectors tend to claim the other sectors when results are unexpectedly poor. Without trust and reliable partners, technical and economic chain results will remain poor. But truthful information exchange between sectors will result in decreased economic losses and improve the total poultry chain performance.

References

Amraei, S., Mehdizadeh, S.A., Sallary, S., 2017. Application of computer vision and support vector regression for weight prediction of live broiler chicken. Engineering in Agriculture Environment and Food 10: 266-271.

Appleby, M.C., Hughes, B.O., 1991. Welfare of laying hens in cages and alternative systems: environmental, physical and behavioural aspects. World's Poultry Science Journal 47: 109-128.

Astill, J., Fraser, E., Dara, R., Sharif, S., 2018. Detecting and predicting emerging disease in poultry with the implementation of new technologies and big data: a focus on avian influenza virus. Frontiers in Veterinary Science 5: 263.

Aydin, A., 2017a. Development of an early detection system for lameness of broilers using computer vision. Computers and Electronics in Agriculture 136: 140-146.

Aydin, A., 2017b. Using 3D vision camera system to automatically assess the level of inactivity in broiler chickens. Computers and Electronics in Agriculture 135: 4-10.

Aydin, A., Bahr, C., Berckmans, D., 2015. Automatic classification of measures of lying to assess the lameness of broilers. Animal Welfare 24: 335-343.

Aydin, A., Berckmans, D., 2016. Using sound technology to automatically detect the short-term feeding behaviours of broiler chickens. Computers and Electronics in Agriculture 121: 25-31.

Bamelis, F., Kemps, B., Mertens, K., De Ketelaere, B., Decuypere, E., DeBaerdemaeker, J., 2005. An automatic monitoring of the hatching process based on the noise of the hatching chicks. Poultry Science 84: 1101-1107.

Banhazi, T.M., Lehr, H., Black, J.L., Crabtree, H., Schofield, P., Tscharke, M., Berckmans, D., 2012a. Precision livestock farming: an international review of scientific and commercial aspects. International Journal of Agricultural and Biological Engineering 5: 1-9.

Banhazi, TM, Babinszky, L., Halas, V., Tscharke, M., 2012b. Precision livestock farming: precision feeding technologies and sustainable livestock production. International Journal of Agricultural and Biological Engineering 5: 54-61.

Ben Sassi, N., Averos, X., Estevez, I., 2016. Technology and poultry welfare. Animals 6: 62.

Bennett, C.E., Thomas, R., Williams, M., Zalasiewicz, J., Edgeworth, M., Miller, H., Coles, B., Foster, A., Burton, E.J., Marume, U., 2018. The broiler chicken as a signal of a human reconfigured biosphere. Royal Society Open Science 5: 180325.

Berckmans, D., 2017. General introduction to precision livestock farming. Animal Frontiers 7: 6-11.

Bessei, W., 2016. Welfare of broilers: a review. World's Poultry Science Journal 62: 455.

Bottje, W.G., Harrison, P.C., Grishaw, D., 1983. Effect of an acute heat stress on blood flow in the artery of Hubbard cockerels. Poultry Science 62: 386-387.

Carroll, B.T., Anderson, D.V., Daley, W., Harbert, S., Britton, D.F., Jackwood, M.W., 2014. Detecting symptoms of diseases in poultry through audio signal processing. In: Proceedings of the IEEE Global Conference on Signal and Information Processing (GlobalSIP), Atlanta, Georgia, USA.

Carvalho, T.M.R., Massari, J.M., Sabino, L.A., Moura, D.J., 2013. Sensor placement to reach thermal comfort and air quality in broiler housing. In: Proceedings of the Precision Livestock Farming 2013 – 6th European Conference on Precision Livestock Farming, ECPLF 2013, Leuven, Belgium, 10-12 September 2013, pp. 945-952.

Carvalho, V.R.R., Bucklin, R.A., Shearer, J.K., Shearer, L., 2005. Effects of trimming on dairy cattle hoof weight bearing and pressure distributions during the stance phase. Transactions of the ASAE 48: 1653-1659.

Chen, Y., Ni, J.Q., Diehl, C.A., Heber, A.J., Bogan, B.W., Chai, L.L., 2010. Large scale application of vibration sensors for fan monitoring at commercial layer hen houses. Sensors 10: 11590-604.

Chien, Y.R., Chen, Y.X., 2018. An RFID-based smart nest box: an experimental study of laying performance and behaviour of individual hens. Sensors 18: 859.

Coelho, D.J.D.R., Ilda De Fátima, F., Baptista, F.J., Souza, C.D.F., De Sousa, F.C., 2016. Mapping the thermal comfort index in laying hens facilities. CIGR-AgEng conference Jun. 26-29, 2016, Aarhus, Denmark. Available at: https://tinyurl.com/yx9mnxsk.

Collins, L.M., 2008. Non-intrusive tracking of commercial broiler chickens *in situ* at different stocking densities. Applied Animal Behaviour Science 112: 94-105.

Corkery, G., Ward, S., Kenny, C., Hemmingway, P., 2013. Incorporating smart sensing technologies into the poultry industry. Journal of World's Poultry Research 3: 106-128.

Curi, T.M.R.C., Conti, D., Vercellino, R.A., Massari, J.M., De Moura, D.J., De Souza, Z.M., Montanari, R., 2017. Positioning of sensors for control of ventilation systems in broiler houses: A case study. Scientia Agricola 74: 101-109.

Curto, F.P.F., Nääs, I.A., Pereira, D.F., Salgado, D.D., Murayama, M., Behrens, F., 2002. Predicting broiler breeder's behaviour using electronic identification. Agricultural Engineering International 4: 323-329.

D'Alfonso, T.H., Manbeck, H.B., Roush, W.B., 1996. A case study of temperature uniformity in three laying hen production buildings. Transactions of the American Society of Agricultural Engineers 3: 669-675.

Dawkins, M.S., Cain, R., Merelie, K., Roberts, S.J., 2013. In search of the behavioural correlates of optical flow patterns in the automated assessment of broiler chicken welfare. Applied Animal Behaviour Science 145: 44-50.

De Jong, I.C., Guémené, D., 2011. Major welfare issues of broiler breeders. World's Poultry Science Journal 67: 73-81.

De Montis, A., Pinna, A., Barra, M., Vranken, E., 2013. Analysis of poultry eating and drinking behaviour by software eYeNamic. Journal of Agricultural Engineering 44: 166-172.

Decuypere, E., Tona, K., Bruggeman, V., Bamelis, E., 2001. The day-old chick: a crucial hinge between breeders and broilers. World's Poultry Science Journal 57: 127-138.

Dominiak, K.N., Kristensen, A.R., 2017. Prioritizing alarms from sensor-based detection models in livestock production-a review on model performance and alarm reducing methods. Computers and Electronics in Agriculture 133: 46-67.

Du, X., Lao, F., Teng, G., 2018. A sound source localization analytical method for monitoring the abnormal night vocalizations of poultry. Sensors 18: 2906.

Duncan, I.J.H., 2001. The pros and cons of cages. World's Poultry Science Journal 57: 381-390.

Eijsakkers, H., Scholten, M., 2011. Over zorgvuldige veehouderij: veel instrumenten, één concert. Wageningen University, Lelystad, the Netherlands.

Exadaktylos, V., Silva, M., Berckmans, D., 2011. Real-time analysis of chicken embryo sounds to monitor different incubation stages. Computers and Electronics in Agriculture 75: 321-326.

Fernandez, A.P., Norton, T., Tullo, E., Van Hertem, T., Youssef, A., Exadaktylos, V., Vranken, E., Guarino, M., Berckmans, D., 2018. Real-time monitoring of broiler flock's welfare status using camera-based technology. Biosystems Engineering 173: 103-114.

Freire, R., Cowling, A., 2013. The welfare of laying hens in conventional cages and alternative systems: first steps towards a quantitative comparison. Animal Welfare 22: 57-65.

Girard, M.T.E., Zuidhof, M.J., Bench, C.J., 2017a. Feeding, foraging, and feather pecking behaviours in precision-fed and skip-a-day-fed broiler breeder pullets. Applied Animal Behaviour Science 188: 42-49.

Girard, T.E., Zuidhof, M.J., Bench, C.J., 2017b. Aggression and social rank fluctuations in precision-fed and skip-a-day-fed broiler breeder pullets. Applied Animal Behaviour Science 187: 38-44.

Heerkens, J.L.T., Delezie, E., Ampe, B., Rodenburg, T.B., Tuyttens, F.A.M., 2016b. Ramps and hybrid effects on keel bone and foot pad disorders in modified aviaries for laying hens. Poultry Science 95: 2479-2488.

Heerkens, J.L.T., Delezie, E., Kempen, I., Zoons, J., Ampe, B., Rodenburg, T.B, Tuyttens, F.A.M., 2015. Specific characteristics of the aviary housing system affect plumage condition, mortality and production in laying hens. Poultry Science 94: 2008-2017.

Heerkens, J.L.T., Delezie, E., Rodenburg, T.B., Kempen, I., Zoons, J., Ampe, B., Tuyttens, F.A.M., 2016a. Risk factors associated with keel bone and foot pad disorders in laying hens housed in aviary systems. Poultry Science 95: 482-488.

Hein, T., 2007. Cutting losses due to cracking and breaking. Poultry World 5: 16-17.

Hulet, R.M., 2007. Managing incubation: Where are we and why? Poultry Science 86: 1017-1019.

Hyun, H.Y., Yeong, H.B., Wongi, M., 2007. Implantable wireless sensor network to monitor the deep body temperature of broilers. In: Proceedings of the 5th ACIS International Conference on Software Engineering Research, Management and Applications, p. 513-517.

Ji, B., Zheng, W., Gates, R.S., Green, A.R., 2019. Design and performance evaluation of the upgraded portable monitoring unit for air quality in animal housing. Computers and Electronics in Agriculture 124: 132-140.

Johnson, R., 2019. McDonald's and FFAR partner up to improve broiler welfare. Available at: https://thepoultrysite.com/news/2019/04/mcdonalds-and-ffar-partner-up-to-improve-broiler-welfare

Kebeler, T., Schiefer, G., 1996. Computer aided environmental control to support environmental-management-systems in agricultural and food-industrial production-chains. University Bonn, Department of Agricultural Economics, Bonn, Germany.

Knowles, T.G., Kestin, S.C., Haslam, S.M., Brown, S.N., Green, L.E., Butterworth, A., Pope, S.J., Pfeiffer, D., Nicol, C.J., 2008. Leg disorders in broiler chickens: prevalence, risk factors and prevention. PLoS ONE 3: e1545.

Kristensen, H.H., Aerts, J.M., Leroy, T., Wathes, C.M., Berckmans, D., 2006. Modelling the dynamic activity of broiler chickens in response to step-wise changes in light intensity. Applied Animal Behaviour Science 101: 125-143.

Lee, J., Noh, B., Jang, S., Park, D., Chung, Y., Chang, H.H., 2015. Stress detection and classification of laying hens by sound analysis. Asian-Australasian Journal of Animal Sciences 28: 592-598.

Li, C., Chang, J., Cheng, C., Hsieh, L., 2011. A novel non-destructive technology for inspecting eggshell cracks using impulsive response time. Food Science and Technology Research 17: 1-10.

Liu, Z., Wang, L., Beasley, D., Oviedo, E., 2007. Effect of moisture content on ammonia emissions from broiler litter: a laboratory study. Journal of Atmospheric Chemistry 58: 41-53.

Lokhorst, C., Ipema, A.H., 2010. Precision livestock farming for operational management support in livestock production chains. In: Trienekens, J., Top, J., Van der Vorst, J., Beulens, A. (eds.) Towards effective food chains: models and applications. Wageningen Academic Publishers, Wageningen, the Netherlands, pp. 293-308.

Lourens, A., 2001. The importance of air velocity in incubation. World Poultry 17: 29-30.

Lourens, A., Jansman, A., Rebel, J., van Harn, J., Veldkamp, T., Stockhofe-Zurwieden, N., Melchior, M., van Emous, R.A., Kense, M., 2011. Verminderen antibioticagebruik in de vleeskuikensector – CLEAR Helpdeskvraag 2011. Report 512. Animal Sciences Group, Wageningen University, Wageningen, the Netherlands.

Lourens, A., Steentjes, A., 2008. Zootechnical and veterinairy factors at breeder level: effects on broiler mortality. Report 194; Animal Sciences Group, Wageningen University, Wageningen, the Netherlands.

Lourens, A., van den Brand, H., Heetkamp, M.J.W., Meijerhof, R., Kemp, B., 2007. Effects of eggshell temperature and oxygen concentration on embryo growth and metabolism during incubation. Poultry Science 86: 2194-2199.

Lourens, A., van den Brand, H., Meijerhof, R., Kemp., B., 2005. Effect of eggshell temperature during incubation on embryo development, hatchability, and posthatch development. Poultry Science 84: 914-920.

McBratney, A., Bouma, J., Whelan, B., Ancev, T., 2005. Future directions of precision agriculture. Springer, Dordrecht, the Netherlands, pp. 7-23.

McGrath, N., Dunlop, R., Dwyer, C., Burman, O., Phillips, C.J., 2017. Hens vary their vocal repertoire and structure when anticipating different types of reward. Animal Behavior 130: 79-96.

McKay, J.C., 2008. The genetics of modern commercial poultry. In: Proceedings of the 23rd World's Poultry Congress, Brisbane, Australia, July 2008. CD-ROM.

MEG, 2018. Marktbilanz Eier und Geflügel 2018. Eugen Ulmer KG Verlag, Stuttgart, Germany.

Mehdizadeh, S.A., Neves, D.P., Tscharke, M., Nääs, I.A., Banhazi, T.M., 2015. Image analysis method to evaluate beak and head motion of broiler chickens during feeding. Computers and Electronics in Agriculture 114: 88-95.

Miragliotta, M.Y., Naas, I.A., Manzione, R.L.E., Nascimento, F., 2006. Spatial analysis of stress conditions inside broiler house under tunnel ventilation. Scientia Agricola 63: 426-432.

Mollo, M.N., Vendrametto, O., Okano, M.T., 2009. Precision livestock tools to improve products and processes in broiler production: A review. Revista Brasileira de Ciência Avícola 11: 211-218.

Mortensen, A.K., Lisouski, P., Ahrendt, P., 2016. Weight prediction of broiler chickens using 3D computer vision. Computers and Electronics in Agriculture 123: 319-326.

Moura, D.J., Nääs, I.A., Alves, E.C.S., Carvalho, T.M.R., Vale, M.M., Lima, K.A.O., 2008. Noise analysis to evaluate chick thermal comfort. Scientia Agricola 65: 438-443.

Nääs, I.A., Laganá, M., Mollo, N.M., 2009. Estimating broiler breeder welfare using image analysis. In: Proceedings of 7th World Congress of Computers in Agriculture and Natural Resources 2009, Reno, Nevada, pp. 22-24.

Nääs, I.A., Sonoda, L.T., Romanini, C.E.B., Morello, G.M., Neves, H.A.F., Baracho, M.S., Souza, S.R.L., Mollo, M.N., Menezes, A.G., Moura, D.J., Almeida, P., 2008. Morphological asymmetry and broiler welfare. Brazilian Journal of Poultry Science 10: 203-207.

Nakarmi, A.D., Tang, L., Xin, H., 2014. Automated tracking and behaviour quantification of laying hens using 3D computer vision and radio frequency identification technologies. Transactions of the ASABE 57: 1455-1472.

Neethirajan, S., 2017. Recent advances in wearable sensors for animal health management. Sensing and Bio-Sensing Research 12: 15-29.

Norton, T., Vranken, E., Exadaktylos, V., Berckmans, D., Lehr, H., Vessier, I., Blokhuis, H., Berckmans, D., 2016. Implementation of precision livestock farming (PLF) technology on EU farms: results from the EU-PLF project. In: Proceedings of the CIGR-AgEng Conference, Aarhus, Denmark, 26-29 June 2016, pp. 1-7.

OECD, n.d. Meat consumption. Available at: https://data.oecd.org/agroutput/meatconsumption.htm.

Oviedo-Rondón, E.O., 2007. Predisposing factors that affect walking ability in turkeys and broilers Available at: https://tinyurl.com/2rsdukhv.

Pereira, D.F., 2003. Avaliação do comportamento individual de matrizespesadas (frango de corte) em função do ambiente e identificaçãoda temperatura crítica máxima. PhD Thesis. FEAGRI-UNICAMP, Campinas, SP, Brazil.

Pereira, D.F., Miyamoto, B.C.B., Maia, G.D.N., Tatiana Sales, G., Magalhães, M.M., Gates, R.S., 2013. Machine vision to identify broiler breeder behaviour. Computers and Electronics in Agriculture 99: 194-199.

Pereira, D.F., Nääs, I.D.A., Gabriel Filho, L.R.A., Neto, M.M., 2012. Cluster index for accessing thermal comfort for broiler breeders. In: Proceedings of the ASABE – 9th International Livestock Environment Symposium 2012, Valencia, Spain, pp. 207-212.

Ren, G., Lin, T., Ying, Y., Chowdhary, G., Ting, K.C., 2020. Agricultural robotics research applicable to poultry production: a review. Computers and Electronics in Agriculture 169: 105216. https://doi.org/10.1016/j.compag.2020.105216

Rizwan, M., Carroll, B.T., Anderson, D.V., Daley, W., Harbert, S., Britton, D.F., Jackwood, M.W., 2016. Identifying rale sounds in chickens using audio signals for early disease detection in poultry. In: Proceedings of the 2016 IEEE Global Conference on Signal and Information Processing (GlobalSIP), Washington, DC, USA, 7-9 December 2016, pp. 55-59.

Rodenburg, T.B., De Reu, K., Tuyttens, F.A.M., 2012. Performance, welfare, health and hygiene of laying hens in non-cage systems in comparison with cage systems. In: Sandilands, V., Hocking, P.M. (eds.) Poultry Science Symposium Series, Volume 30, CABI, Glasgow, pp. 210-224.

Romanini, C.E., Exadaktylos, V., Tong, Q., McGonnel, I., Demmers, T.G., Bergoug, H., Eterradossi, N., Roulston, N., Garain, P., Bahr, C., Berckmans, D., 2013. Monitoring the hatch time of individual chicken embryos. Poultry Science 92: 303-309.

Roush, W.B., Wideman Jr., R.F., 2000. Evaluation of broiler growth velocity and acceleration in relation to pulmonary hypertension syndrome. Poultry Science 79: 180-191.

Rowe, E., Dawkins, M.S., Gebhardt-Henrich, S.G., 2019. A systematic review of precision livestock farming in the poultry sector: is technology focused on improving bird welfare? Animals 27: 614. https://doi.org/10.3390/ani9090614

Scholten, M.C.T., De Boer, I.J.M., Gremmen, B., Lokhorst, C., 2013. Livestock farming with care: towards sustainable production of animal-source food. NJAS – Wageningen Journal of Life Science 66: 3-5.

Shimmura, T., Hirahara, S., Azuma, T., Suzuki, T., Eguchi, Y., Uetake, K., Tanaka, T., 2010. Multi-factorial investigation of various housing systems for laying hens. British Poultry Science 51: 31-42.

Stevenson, P., 2017. Precision livestock farming: could it drive the livestock sector in the wrong direction. Available at: https://tinyurl.com/ne8jmnrt

Traffano-Schiffo, M.V., Castro-Giraldez, M., Herrero, V., Colom, R.J., Fito, P.J., 2018. Development of a non-destructive detection system of deep pectoral myopathy in poultry by dielectric spectroscopy. Journal of Food Engineering 237: 137-145.

Usher, C.T., Daley, W.D., Joffe, B.P., Muni, A., 2017. Robotics for poultry house management. In: Proceedings of the ASABE Annual International Meeting, Spokane, Washington.

Van Erp-Van der Kooij, E., Rutter, S.M., 2020. Using precision farming to improve animal welfare. CAB Reviews 15: 051. https://doi.org/10.1079/PAVSNNR202015051

Van Hertem, T., Norton, T., Berckmans, D., Vranken, E., 2018. Predicting broiler gait scores from activity monitoring and flock data. Biosystems Engineering 173: 93-102.

Van Horne, P.L.M., 2018. Competitiveness of the EU poultry meat sector, base year 2017; International comparison of production costs. Report 2018-116. Wageningen Economic Research, Wageningen, the Netherlands, 40 pp.

Van Horne, P.L.M., 2019. Competitiveness of the EU egg sector, base year 2017; International comparison of production costs. Report 2019-008. Wageningen Economic Research, Wageningen, the Netherlands, 52 pp.

Vroegindeweij, B.A., Blaauw, S.K., IJsselmuiden, J.M.M., van Henten, E.J., 2018b. Evaluation of the performance of PoultryBot, an autonomous mobile robotic platform for poultry houses. Biosystems Engineering 174: 295-315.

Vroegindeweij, B.A., IJsselmuiden, J.M.M., van Henten, E.J., 2016. Probabilistic localisation in repetitive environments: estimating a robot's position in an aviary poultry house. Computers and Electronics in Agriculture 124: 303-317.

Vroegindeweij, B.A., van Hell, S., IJsselmuiden, J.M.M., van Henten, E.J., 2018a. Object discrimination in poultry housing using spectral reflectivity. Biosystems Engineering 167: 99-113.

Vroegindeweij, B.A., van Willigenburg, G.L., Koerkamp, P.W.G.G., van Henten, E.J., 2014. Path planning for the autonomous collection of eggs on floors. Biosystems Engineering 121: 186-199.

Wathes, C.M., Kristensen, H.H., Aerts, J.M., Berckmans, D., 2008. Is precision livestock farming an engineer's daydream or nightmare, an animal's friend or foe, and a farmer's panacea or pitfall? Computers and Electronics in Agriculture 64: 2-10.

Werner, A., Jarfe, A., Stafford, J.V., Cox, S.W.R., Sidney, W.R., 2003. Programme Book of the Joint Conference of ECPA-ECPLF: 1st European Conference on Precision Livestock Farming and 4th European Conference on Precision Agriculture; Wageningen Academic Publishers, Wageningen, the Netherlands, 2003. Available at: https://tinyurl.com/2paafxwa.

Whitaker, B., Carroll, B., Daley, W., Anderson, D., 2014. Sparse decomposition of audio spectrograms for automated disease detection in chickens. In: 2014 IEEE Global Conference on Signal and Information Processing (GlobalSIP), pp. 1122-1126.

Wilhelmsson, S., Yngvesson, J., Jönsson, L., Gunnarsson, S., Wallenbeck, A., 2019. Welfare Quality® assessment of a fast-growing and a slower-growing broiler hybrid, reared until 10 weeks and fed a low-protein, high protein or mussel-meal diet. Livestock Science 219: 71-79.

Winckler, C., 2019. Assessing animal welfare at the farm level: Do we care sufficiently about the individual? Animal Welfare 28: 77-82.

Woodcock, M.B., Pajor, E.A., Latour, M.A., 2004. The effects of hen vocalizations on chick feeding behaviour. Poultry Science 83: 1940-1943.

Yanagi, J.R.T., Xin H., Gates R.S., 2002. A research facility for studying poultry responses to heat stress and its relief. Applied Engineering in Agriculture 18: 255-260.

Zaninelli, M., Redaelli, V., Luzi, F., Bontempo, V., Dell'Orto, V., Savoini, G., 2017. A monitoring system for laying hens that uses a detection sensor based on infrared technology and image pattern recognition. Sensors 17: 1195.

Zaninelli, M., Redaelli, V., Tirloni, E., Bernardi, C., Dell'Orto, V., Savoini, G., 2016. First results of a detection sensor for the monitoring of laying hens reared in a commercial organic egg production farm based on the use of infrared technology. Sensors 16: 1757.

Zhang, F., Hu, Y., Chen, L., Guo, L., Duan, W., Wang, L., 2016. Monitoring behaviour of poultry based on RFID radio frequency network. International Journal of Agricultural and Biological Engineering 9: 139-147.

Zhao, Y., Aarnink, A.J.A., Hofschreuder, P., Groot Koerkamp, P.W.G., 2009. Evaluation of an impaction and a cyclone pre-separator for sampling high PM10 and PM2.5 concentrations in livestock houses. Journal of Aerosol Science 40: 868-878.

Zhuang, X., Bi, M., Guo, J., Wu, S., Zhang, T., 2018. Development of an early warning algorithm to detect sick broilers. Computers and Electronics in Agriculture 144: 102-113.

Zuidhof, M.J., 2018. Lifetime productivity of conventionally and precision-fed broiler breeders. Poultry Science 97: 3921-3937.

Zuidhof, M.J., Fedorak, M.V., Ouellette, C.A., Wenger, I.I., 2017. Precision feeding: Innovative management of broiler breeder feed intake and flock uniformity. Poultry Science 96: 2254-2263.

Zuidhof, M.J., Schneider, B.L., Carney, V.L., Korver, D.R., Robinson, F.E., 2014. Growth, efficiency, and yield of commercial broilers from 1957, 1978, and 2005. Poultry Science 93: 2970-2982.

6. Sensors and automated monitoring in horses

E.K. Visser[1*] and M. de Kort[2]

[1]Aeres University of Applied Sciences, P.O. Box 374, 8250 AJ Dronten, the Netherlands; k.visser@aeres.nl
[2]HAS University of Applied Sciences, P.O. Box 90108, 5200 MA 's Hertogenbosch, the Netherlands

Highlights

- It is estimated that there are about 7 million horses in Europe of which 450,000 in the Netherlands.

- The role of the horse in developed countries is very different from the role of horses in developing countries in which they are solely for horsepower in agriculture and transport.

- The horse-human dyad is unique in sports and leisure and therefore there is much interest and debate on how to guarantee a horse's welfare.

- There is a large variety of commercial sensors and wearables to monitor the health and performance of horses.

- Only recently the sensor market is also focusing on the added value of sensors in maintaining horse welfare.

- Commercial devices range from measuring physical aspects like rein tension, physiological parameters such as heart rate to behavioural aspects like activity patterns and posture.

- There is an increasing interest in small, light and flexible devices to not hamper the horse and rider in their activities.

6.1 Background of horses sector

Horses play an enormous role in human societies worldwide, with uses in leisure activities, sports and for working purposes. Equestrian sports, such as show jumping and dressage, focus on the level of control and balance between horse and rider, while working roles such as mounted police units, search and rescue teams and therapy settings focus more on the relationship and confidence between horse and human. Foremost the largest group of equines (80%) lives in developing countries in which horses are used for horsepower in agriculture and transport (Waran, 2002). Currently, there are over 300 breeds of horses worldwide. The best dressage, showjumping and eventing horses in the world are bred in the Netherlands, as evidenced by the World Breeding Federation for Sport Horses (WBFSH) ranking, where the KWPN was again the number one studbook for dressage, jumping and eventing in 2020 (www.wbfsh.org).

6.1.1 Number of horses and horse enthusiasts

It has been estimated that there are 80 million riders worldwide and more than 60 million horses. The worldwide economic value of the international equestrian sector is unknown. In 2005, the economic value of the US equestrian industry was estimated at $200 billion. The EU equine sector employs at least 896,000 persons across the EU; is worth over 100 billion Euro per year and uses at least 2.6 million hectares of land in the EU (World Horse Welfare, 2015). Even though the European Commission regulations require the identification of kept *Equidae*, nobody is sure exactly how many horses live in the fields, meadows, paddocks and barns of the EU. The World Horse Welfare charity estimates that there are approximately 7 million horses in Europe. However, member states have provided very different figures per country. Taking the means of these different figures per country, in 2015 the country with the most horses was estimated to be France with 840,000 horses, followed closely by the UK with 796,000 horses, and Romania with 728,814 horses. Germany comes in place six with 480,500 horses and the Netherlands in 8th place with a mean of 293,500 horses (World Horse Welfare, 2015). Across Europe, the position of *Equidae* is constantly changing. In some areas, they have moved away from their roots as working animals, and are now predominantly used for leisure and sports, while in other parts of Europe they are still very much functional animals, used for semi-subsistence farming, tourism and food production (Figure 6.1).

6. Sensors and automated monitoring in horses

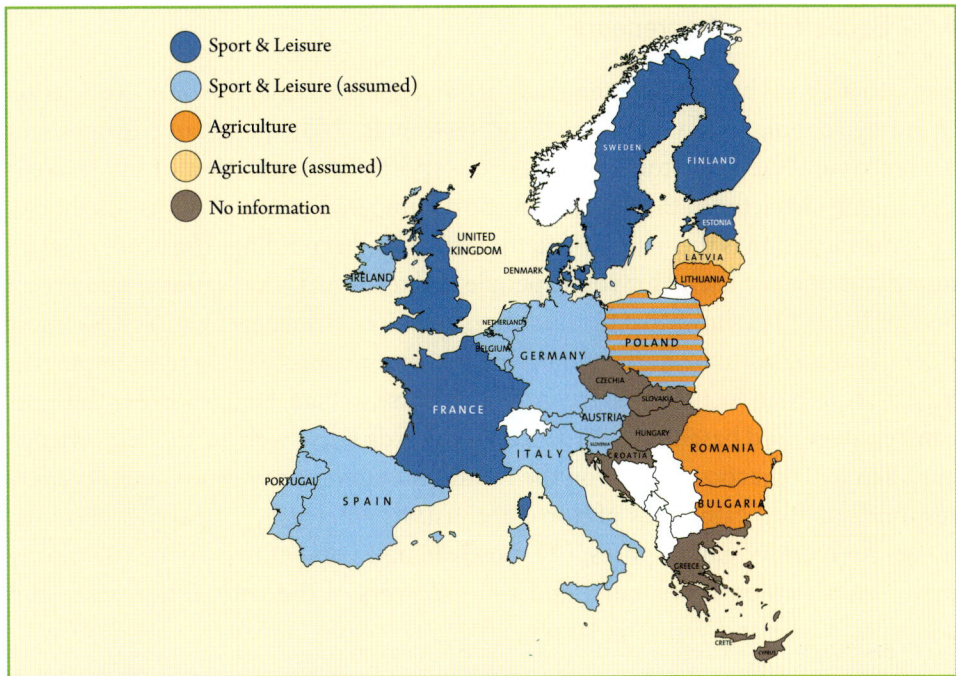

Figure 6.1. Role of horses in Europe (courtesy World Horse Welfare, www.worldhorsewelfare.org).

The figures for the Netherlands range from 137,000 to 450,000 horses. The most reliable source is of the Dutch Horse Sport Organization (KNHS) which estimates 450,000 horses in the Netherlands (KNHS, 2018). Using the mean equine population figures, Belgium has the highest per capita equine population with approximately one equine per 21 persons, followed by Romania and Ireland. The Netherlands comes in sixth place with one equine per 59 persons; however, when the figures of the KNHS are used (450,000 horses) the Netherlands has one equine per 38 persons (World Horse Welfare, 2015).

Horses have an important role in Dutch society. In 2016, it was estimated that there were approximately 450,000 horses in the Netherlands and 400,000 people sporting with horses. The turnover was estimated to lie between 1.5 en 2 billion euro. Over 75,000 of those people riding horses do compete. Over 1.2 million people in the Netherlands watch horse sports. There are approximately 400,000 active riders in the Netherlands, of which are approximately 200,000 members of the Dutch Horse Sport Organization (KNHS). Although the horse entrepreneurs are predominantly men, 80% of all equestrian athletes are women (HAS Hogeschool, 2016; KNHS, 2018; Visser and Van Wijk-Jansen, 2012).

6.1.2 Equestrian entrepreneurs

Equestrian sports have undergone a radical transition in recent years, from association to riding school, from volunteers to professionals. The equestrian entrepreneur has become the cornerstone of equestrian sports. The attraction of the horse is the unique character of the equestrian sport compared to other sports: the horse-human interaction is the main reason to start and to keep riding. The horse plays an important role as a friend for young people and teenagers. The National Equestrian Survey 2014/2015 shows that the rider-horse relationship changes with age. For most riders, bonding with the horse and having fun with the horse is the most important aspect of this relationship. With the interest in horse welfare, Dutch horse enthusiasts can be categorised based on their attitudes towards horse welfare, their knowledge and their daily practices. Based on the results of a survey with approximately 2,500 respondents four categories of horse enthusiasts emerged: (1) people who own horses for recreational purposes; (2) mostly female competitive riders; (3) mostly male equestrian entrepreneurs; and (4) followers of the natural horsemanship concepts (Visser and Van Wijk-Jansen, 2012).

There is a wide variety of professions and businesses all working in the equine industry: farriers, veterinarians, riding instructors, grooms, riding equipment retailers, construction of facilities, livery yards, transporters and so on. The Netherlands has over 10,000 equine-related companies and 3,000 equestrian sports centres and associations. In recent years turnover, margins and profits are increasing. Also, there seems to be enough confidence in the future to choose to invest in the company. These developments create confidence in the future of the primary companies in the equestrian industry (HAS Hogeschool, 2019).

6.1.3 Trends and concerns

Horse welfare is currently the subject of debate within the equine industry. The welfare of horses is important as a showcase for the sector. The size of the companies is increasing, making profitable business operations increasingly important. We see the industry becoming more professional. Due to the unique aspect of a sport involving a living creature with no voice of its own, a horse's welfare is dependent upon the discretion of its rider. However, since we are not able to directly communicate with animals, decisions regarding their care are complex and difficult. Often decisions are not based solely on science as the interests and values of all stakeholders must be considered and, in some cases, insufficient empirical evidence exists to even accurately decide whether a practice is right or wrong (World Horse Welfare, 2015).

Welfare concerns are related to: (1) problems with the environment and ways in which horses are kept: environments with lack of space, long periods of confinement and restricted possibilities for social interactions; (2) feeding management: the amounts roughage versus concentrates and especially the time between meals; resulting in gastric ulcers and obesity; and (3) training methods: lack of understanding, knowledge and expertise to practice learning theories correctly. Consequently, the use of bits, spurs and harness causing discomfort or pain for horses. Besides these content related welfare concerns, an overarching concern for welfare problems in horses is the growing number of horse enthusiasts lacking the appropriate knowledge and skills and not being able to accurately read and interpret horse signals and cues indicating stress symptoms (Figure 6.2).

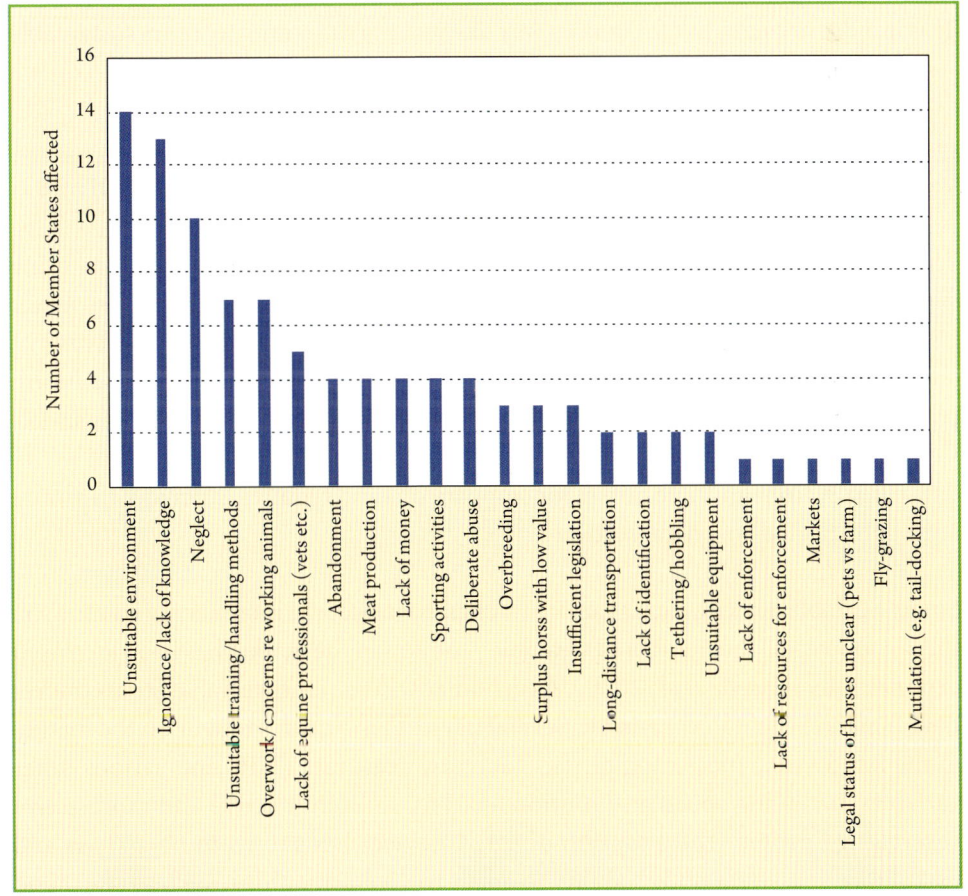

Figure 6.2. Prevalent welfare problems in Europe reported by NGO's (Courtesy World Horse Welfare, www.worldhorsewelfare.org).

The training and management of horses is quickly evolving towards a more tailored practice, where the individual needs of horses are met reflecting their unique physiology and behaviour. The increased demand and further development of new sensor technologies for horses is a direct result of the increased awareness of the horse owner in equine welfare and in the need to adapt training and management to improve the overall welfare of each horse. Particularly in the developed countries of the world, animal welfare and the public discussion associated with it are becoming increasingly important (Waran *et al.*, 2002).

Another debate currently in the equine industry is the Social License to Operate. Where and how do we want to ride horses, organise competitions and house our horses in the future? In its report 'Horsemanship Competence', FEI cite the growing criticism of equestrian sports as one of the trends that 'the sector' should be aware of. This trend is not new, but for many – just like making the company more sustainable – still a far-from-my-bed show (Krysiak, 2017).

6.2 Application of sensors and automated monitoring in horses

The ability to monitor vital signs over long periods and during exercise can bring significant benefits to training horses both in terms of better performances, fewer risks of injury and improved welfare through the early detection of disease or abnormalities in daily behavioural patterns. The idea of monitoring horse health conditions isn't new but it's normally very expensive and done by veterinarians. Collecting and interpreting results is complex and involves the processing of lots of data (usually from multiple horses for comparison) and requires expensive hardware and sophisticated analytical software.

Smart wearable devices, such as smartwatches, have become increasingly common in the last few years. The number of wearable products in the market has been increased exponentially (Casella *et al.*, 2020). As in humans, the equine wearable fitness tracker device market is in full expansion. Several devices have been brought on the market in the last 5 years, measuring physical activity, heart rate (HR) and various other variables, including speed and motion analysis, in both inactive and more active (exercising) people. The adoption of smart devices has shown to be an easy and useful way to assess a body's ability to adapt to given situations, such as short and long term training or rehabilitation and, as such, to regularly monitor fitness (Ter Woort *et al.*, 2021).

With the use of wearables in the horse industry, variables such as activity, posture, heart rate (variability), respiratory rate and temperature can provide immediate and accessible data to owners, trainers veterinarians, and other stakeholders during all aspects of a horse's day-to-day activity including stabling, exercise, travel, and veterinary practices. In the last decade monitoring health and welfare in horses has gained attention especially in the field of equestrian sports where horse and rider interact and responses of horses are helpful to rider and trainer for performance, health and welfare concerns. An increasing number of riders and Olympic Teams implement monitoring heart rate or rein tension measurements in the preparation to competition to optimise the training (https://www.moxiesport.nl and https://nl.ipostechnology.com). In 2020 and 2021 when the COVID19 pandemic forced livery yards over the whole world to close for horse owners, there was, especially in the USA, a growing interest in monitoring customers horses' welfare and health with camera's and wearables.

In equitation science, which specifically aims to promote and encourage the application of objective research and advanced practice to improve the welfare of horses, a wide range of measurements can be taken from the horse and/or rider, including physical data such as pressure and tension, physiological measures such as heart rate and temperature and behavioural measures like activity and posture. To monitor health, welfare and performance in horses either in their home environments as well as during training and competition a range of technologies have been developed. The range of different techniques and equipment are both custom made and commercially available. The list of wearables mentioned in Table 6.1 is by no means exhaustive, only commercially available sensors have been included and wearables are not exclusively dedicated to measuring solitary performance or health or welfare; on the contrary, measures of physical, physiological or behavioural responses all contribute to an overall assessment of performance, health and welfare.

6.2.1 Physical measures

Pressure applied to the horse body is an everyday practice in horse training. Force applied to a small area can result in very high pressure. Tack is designed to either disperse pressure (e.g. in the case of the saddle) or concentrate pressure (e.g. in the case of head restraint). It has been suggested that in tack designed to concentrate pressure, the smaller the interface with the horse, the more severe the device and the greater the need for excellent timing in pressure-release (McGreevy et al., 2018).

Table 6.1. Overview of sensor used for horses, in development or commercially available.

PLF sensor	Where	What it measures	Why	Company website
Pressure sensor	legs	load	health, training	ekico.fr
	saddle pad	saddle pressure	health, training	estride.store; novelelectronics.de
	reins	rein tension	health, training, stress	ipostechnology.com
	between hoof and shoe	gait analysis	health, training	tekscan.com
	legs	blood pressure	health	ponyuptechnologies.com
Heart rate sensor	girth attachment	heart rate	health, training, stress	polar.com; arioneo.com; equisense.com; piavita.com; trackener.com
	headpiece of the bridle	heart rate	health, training, stress	equ.la
	girth attachment	heart rate	health, training	seaverhorse.com
	legs	heart rate	health, training	ponyuptechnologies.com
Temperature sensor	girth attachment	surface temperature	health	piavita.com
	ingestible pill	core temperature	health	mytemp.eu; hqinc.net
	rug attachment	temperature under rug	health	arioneo.com
	horses' mane	ambient temperature	health, training	steedwatch.com
Resistance	girth attachment	respiration rate	health, training	piavita.com
Accelerometer	rug attachment	activity and posture	health	arioneo.com
	headpiece of the bridle	head angle	training	equ.la
	halter	activity and posture	health	nightwatch24.com
	girth attachment	activity	health, training	piavita.com; seaverhorse.com; trackener.com; equisense.com
	horses' mane	activity	health, training	steedwatch.com
	markers on body	gait analysis	health, training	equimoves.nl
Positioning	rug attachment	distance, speed, position	training	arioneo.com; equisense.com
	headpiece of the bridle	distance, speed, position	training	equ.la
	girth attachment	distance, speed, position	training	seaverhorse.com; trackener.com
	horses' mane	distance, speed, position	health, training	steedwatch.com
Vision	camera	gait analysis	health, training	qualisys.com; simi.com
		surface temperature	health, training, stress	satir.com

Saddle pressure

Most horses that are being ridden have some form of 'saddle'. A saddle is constructed around a wooden tree structure distributing the weight of the load of the rider. Modern alternatives include semi-rigid and treeless saddle designs (Fruehwirth et al., 2004). Good saddle fit enhances equine performance whereas an ill-fitting saddle may be associated with back pain, a poor attitude to work and reduced performance (Clayton et al., 2014). Awareness of the implications of saddle fitting for equine health, performance and welfare has stimulated the development of novel saddle designs. Traditionally, saddle fit has been assessed subjectively by visual and palpatory evaluation under static conditions, which does not take into account changes in the horse's back shape during locomotion or the effects of the rider's weight (Clayton et al., 2014). The most popular system adopted by equine researchers uses a capacitive sensor. When a force changes the relative distance between two closely associated membranes, there is a corresponding change in the capacitance which is converted to an electrical signal that is recorded. The pressure pads produce two measures: Maximum Overall Force (MOF) and Centre of Pressure (COP) with typical MOF values reported of 302 N in walk, 254 N in trot and 172 N in canter. Unfortunately, the use of saddle pads result in an underestimation of the pressure because only perpendicular forces are being recorded (Fruehwirth et al., 2004).

Rein tension

Horses receive signals from the rider's hands via the reins and the bit and equipment worn on the head (e.g. bridle, headcollar). The connection between horse and rider is ideally a light 'physical connection' (Randle et al., 2017). Riders find it difficult to quantify the amount of tension they apply; Clayton et al. (2005) and Randle et al. (2013) reported that riders over-estimated the amount of tension they were applying to the horse. Furthermore, riders' perception of rein tension showed to be more related to the quality (e.g. stability) rather than the quantity or variability of tension (Christensen et al., 2021). Maximum forces that have been found are 43 N in walk, 51 N in trot and 104 Newton in Canter (Clayton et al., 2005). Apart from these measures, at least so important is the consistency of the tension and the speed with which the tension is applied and released. Whilst it is commonly agreed upon that rein tension should be symmetrical and light, a value or range of values that is considered as acceptable and light has yet to be ascertained and agreed (Randle et al., 2017).

6.2.2 Physiological measures

To evaluate vital signs and the (emotional) response of the horse to ridden work, several physiological measures are commonly used: heart rate, temperature and respiration.

Heart rate

Heart rate monitors enable cost-efficient, non-invasive measurement of the cardiovascular function under a range of field conditions including training and competition. Additionally, heart rate (HR) measurements have also been used to evaluate mental states, such as novelty or human-horse interactions during handling (Visser *et al.*, 2002) and horse health. With measuring heart rate parameters (Figure 6.3) the response of the autonomic nervous system and especially the balance between the parasympathetic and sympathetic branch is being monitored with heart rate variability (HRV) (Von Borell *et al.*, 2007). HRV is extensively used in human sports medicine, to assess training, overtraining, fitness and recovery (Dong, 2016). In horses, HRV during exercise has been determined in normal horses (Frick

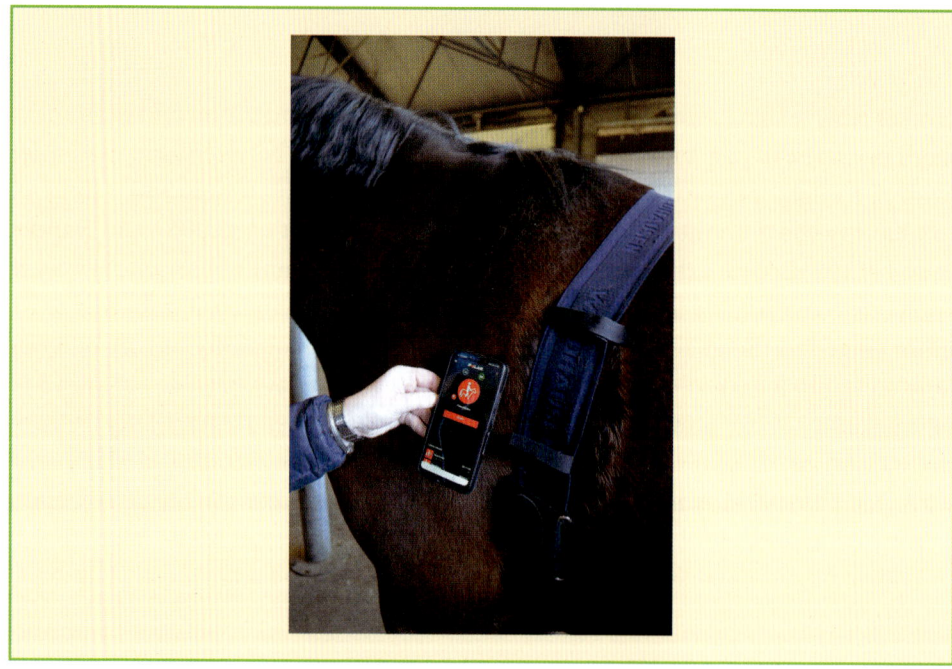

Figure 6.3. Measuring heart rate of both horse and handler (photo: Aeres University of Applied Sciences, Dronten).

et al., 2019) and horses with arrhythmias (Frick et al., 2019). HRV has been used for the detection of heart rate arrhythmia (Broux et al., 2018; Frick et al., 2019). In ill horses, measurement of HRV has been used as a tool to grade pain in cases of laminitis (Rietman et al., 2004) or colic (Graubner et al., 2011). Increasingly, HRV parameters are being used to assess stress levels and monitor welfare in horses (Schmidt et al., 2010).

To be valid, the measurement of HR needs to be assessed in comparison to a golden standard allowing verification of the R-R interval detection. The electrocardiogram (ECG) provides the most accurate measurement of heart activity and is therefore seen as the golden standard for heart rate monitors. Various HR monitors, based on commercially available human devices, have been used to measure HR and derive HRV parameters in equine studies. The Televet ECG system is considered the gold standard for ECG recording at rest and during exercise in horses (ter Woort et al., 2021). Unfortunately, several studies have shown a poor correlation between heart rate variability (HRV) measured by an electrocardiogram (ECG) and the R-R interval provided by commercially available devices, especially during exercise (Ille et al., 2014; Parker et al., 2010; Vendrig, 2013).

In horses, the adoption of heart rate monitors and more sophisticated smart devices has been slower compared to humans. This is probably at least partially related to their lack of practicality. Recently, more complete and user-friendly devices have emerged in the equine market and have been validated at rest (Parker et al., 2010; Ter Woort et al., 2021). Most sensors measuring heart rate use a girth attachment or have the electrodes integrated into an elastic belt (Table 6.1). Some wearables however have the measuring device in a different place (head or legs).

Temperature

The evaluation of body temperature represents a valuable tool to monitor the physiologic status, welfare, and stress response of animals. Similar to humans, skin is the only heat dissipation way of the equine body; when the body energy metabolism and heat production increases, heat dissipation through the skin increases, too, and so the surface temperature rises. Many methods are used to measure body temperature as conventional digital temperature (rectal temperature), infrared thermal image (eye or skin temperature), central venous temperature, percutaneous thermal sensing microchips (PTSM) (muscle temperature), or gastrointestinal pills (gastrointestinal temperature) (Green, 2020). The central venous temperature is seen as the core body temperature (Kang et al., 2020)

Monitoring horse surface temperature allows the evaluation of the function of individual parts of the body (Rizzo et al., 2017). Acute stress responses result in patterns of vasoconstriction and dilation causing changes in underlying blood flow. Changes in underlying blood flow can be detected with changes in surface temperature and these can be measured with infrared thermal imaging cameras. These cameras are being used in medical/veterinary applications to monitor changes in body(parts) surface temperature to detect early, pre-clinical signs of disease.

While some studies have used pressure-sensing mats for the objective evaluation of saddle fit an alternative method for determining the pressure distribution across the saddle can be obtained through the use of infrared thermography (IRT) on the saddle panels (Arruda et al., 2011). To ensure that measures taken using IRT are valid and accurate several factors need to be controlled for. These include ambient conditions, coat colour, surface moisture and distance from which the measures are taken. The accuracy of the thermal cameras available varies considerably and validation against other temperature-based measures is recommended.

Monitoring changes in eye temperature has been applied to assessing the response of horses to potentially stressful situations. During an acute stress response, eye temperature tends to increase, as a result of increased dilation of the blood vessels in the eye. During a potentially aversive procedure like clipping eye temperature increases throughout the procedure, peaking at the end of the ten minutes of clipping and only decreases after the procedure is stopped (Yarnell et al., 2013). Also, the potential discomfort associated with tight nose-bands and head carriage with a nasal plane behind the vertical results in the increase of eye temperature (Fenner et al., 2016). Since the temperature of the eye and surrounding area are less affected by coat variations (colour, thickness, etc.) than the surface temperature, the eye temperature offers the better potential for accurately monitoring emotional responses (Yarnell et al., 2013). Research has shown that there is a relation between increases in core body temperature and eye temperature being not affected by locomotor activity.

Furthermore, thermal sensing microchips may be used to monitor body temperature. The implantation of the percutaneous thermal sensing microchips (PTSMs) internally in the animal is minimally invasive, requiring only the injection of the microchip through a large gauge needle like the conventional ID microchip insertion. After initial implantation, measuring body temperature is completely non-invasive. Robinson et al. (2008) found that percutaneous thermal sensing microchips (PTSMs) were a reliable alternative to rectal thermometry for the measurement of body temperature in horses at rest in an ambient temperature >15.6 °C. Temperatures obtained using PTSMs during exercise have a strong correlation with central venous temperature which is seen as the core body temperature, but show a poor correlation with rectal temperature (Kang et al., 2020).

To remotely monitor the thermal comfort of rug-wearing horses, devices are on the market that measure under-rug temperature (°C) and humidity (%RH), providing a warning when horses become thermally 'uncomfortable'. These wearables are ~3 °C lower than a temperature sensor in direct contact with the horses' coat (Bartlett et al., 2018). With a rug-monitoring device, it was shown that there is a good correlation between the rug weight and temperature under the rug. Based on a handful of studies it is believed that horse owners easily over-rug their horses.

The gastrointestinal temperature (GT) monitoring system comprises ingestible pills (transmitter) and an external recording device (receiver). They have been used for continuous monitoring of thermoregulation in human athletes during field and exercise studies and have been tested in resting and exercising dogs, in elephants, cattle and horses (Verdegaal et al., 2017). The technology is minimally invasive and wireless, and it enables real-time display on a mobile device. A good correlation between gastrointestinal (GI) temperature pill and rectal temperature was found although the GI temperature reached higher maximum temperatures compared to the rectal temperature which may suggest that GI temperature may be more reflective of changes in body temperature during exercise than is the rectal temperature (Green et al., 2005).

Respiration rate

Hot weather, demanding exercise and transport have shown to affect horses' thermoregulation resulting in elevated body temperature and respiration rate. In practice, respiration rate is measured by observing flank movement and recording the elapsed time for 10 full breaths using a stopwatch converting the values subsequently to breaths/min. In experimental settings, mainly exercising on treadmills, respiration rate has been monitored with temperature sensors placed in the trachea and at the entrance to the nostril (Young et al., 1995). More recently, the pattern of breathing can also be monitored by the technique of impedance plethysmography; a non-invasive technique that uses electrical monitoring in the form of resistance. The electrical resistance between electrodes placed on the chest is converted into a voltage signal that varies with the degree of chest expansion. This technique can be incorporated with telemetric ECG monitoring systems using the same electrode positions (Marlin and Roberts, 1998).

6.2.3 Behavioural measures

Observational studies are inherently time and labour intensive. Continuous automated monitoring systems represent an alternative for quantifying the behaviour of horses. Movement sensors or accelerometers offer an objective measurement of mobility. Previous studies have mainly used movement sensors for gait analysis and the detection of lameness and asymmetry in the moving horse. However, these sensors can also be used to monitor habitual behaviours in the paddock or stable, including time spent lying down, the number of lying bouts, and the number of steps taken.

Activity

It is generally agreed that housing conditions, feeding methods and the possibility of free movement have a major influence on the health status and welfare of domesticated horses (Freire *et al.*, 2009). In particular, lack of movement and improper feeding are responsible for diseases of the musculoskeletal system as well as digestive and behavioural disorders. Under natural conditions, horses spend most of their time moving in search of food. Wild-living horses, such as Camargue horses, spend 51 to 64% of the day grazing and 4 to 10% of the day walking (Duncan, 1980).

A handful of studies have explored the use of accelerometry as well as other means of activity tracking in horses (Morrison *et al.*, 2015). Accelerometers have been validated for use in horses exercised in hand (Fries *et al.*, 2017), and for recording activity patterns, such as lying behaviour, in adult horses (DuBois *et al.*, 2015) and foals (Murase *et al.*, 2018). Characterisation of lying behaviour in horses has significant implications for welfare assessments. Specifically, the requirement for horses to achieve lateral recumbency for REM sleep (Hale and Huggins, 1980) may have direct effects on biological functioning. In a study by Murase *et al.* (2018), it was found that foals lie down longer, and more in lateral recumbency, in stables than on pasture. Sleep deprivation may result from chronic pain or environmental insecurity, such as social isolation, causing impairments including severe weight loss or injuries after collapsing (Bertone, 2006). The lying position of the mare in the late stages of gestation is different from normal and this can be detected with a birth alarm. Usually, mares will rest in a half lying or sitting position, predicting the onset of foaling.

Tail movements

In a study by Auclair-Ronzaud *et al.* (2020) the frequency and duration of tail movements were monitored within 24 h of foaling with the use of an accelerometer placed at the base of the tail of the dam using a tail wrap. The tail accelerometer showed that tail movements are more frequent and have shorter durations when approaching foaling.

Head-neck position

The Fédération Equestre Internationale (FEI) has stated in their regulations that:

> The head of the horse should remain in a steady position, as a rule slightly in front of the vertical, with a supple poll as the highest point of the neck, and no resistance should be offered to the Athlete.

At the end of the test, the general impression of the position of the head is evaluated by the judges; a persistent faulty position can have a significant effect on the final score. In studies evaluating the head-neck position of elite dressage horses, frames of video recordings were analysed with commercially available software to measure the angle with the vertical (Lashley *et al.*, 2014). Despite the relevance, the use of an accelerometer to monitor head-neck position continuously has only just recently been applied in practice and is available as a commercial wearable.

Gait analysis

Next to measuring general activity in horses, accelerometers can also be used for more precise measurements for monitoring health issues such as lameness. Lameness is by far the most expensive health issue in the equine field and equine veterinarians spend up to 40% of their working time assessing lameness (Loomans *et al.*, 2007). The gaits of a horse are all characterised by a sequence of hoof steps touching the ground. Currently, most veterinarians rely on subjective visual examination of the gait to detect movement asymmetries that are the common clinical sign of lameness. However, subjective lameness assessment has been shown to have some substantial drawbacks, most due to the limitations of human visual symmetry perception and the bias effect, which ultimately leads to a poor agreement between veterinarians (Arkell *et al.*, 2006). With the use of accelerometers the gait analysis, and thus also the detection of lameness can be measured objectively. For gait analysis different systems are on the market and/or for research purposes; these set-ups include a system with accelerometers per leg and a GPS sensor to account for location. Commercial products for the recognition of horse gaits show limitations in terms of the extra and specialised hardware that is required to classify the performed activities. For gait analysis to detect lameness it is very important to rely upon devices that are validated in scientific studies.

Location

Little is known about the movement patterns of domestic horses or the effects of the various forms of the yard and paddock confinement on distance travelled. Confinement and feeding patterns parallel the modern human lifestyle is thought to be a major contributor to the current high prevalence of obesity, cardiovascular disease, arthritis and diabetes (Hampson *et al.*, 2010). Global positioning systems (GPS) are lightweight and accurate and can measure speed variation and distinguish small movements of horses. GPS has been used to measure locomotion variables in horses during cross-country riding and in foals at pasture (Hebenbrock *et al.*, 2005; Kurvers *et al.*, 2006). Data collected via GPS can be used to characterise individual workloads as speed and distance travelled. GPS data may be paired with supplementary findings from accelerometers, or measures of the internal workload from heart rate monitors (Best and Standing, 2019).

Grazing

Equine grazing studies that incorporate behavioural assessments have traditionally relied on observational methodologies for the evaluation of grazing activity (Weinert *et al.*, 2020). Continuous automated monitoring systems represent an alternate strategy for quantifying the behaviour of horses maintained on pasture, but the application of this approach has seen limited use in equine pasture studies. As for cattle a chewing halter was developed and validated for horses for the detection of jaw movement in stalled horses consuming small meals of concentrate, silage, or hay (Werner *et al.*, 2016). It was found that the total number of bites or chews per day was greater in horses on pasture than when maintained in stables with *ad libitum* hay; whereas the counts for 'other' jaw movements were greater in horses maintained in stables. It has been suggested that an inherent 'ethological' drive for foraging behaviour exists in the equines and this was also suggested by the study of Ellis *et al.* (2006) in which horses offered low fibre diets recorded more time spent chewing on bedding materials compared to horses offered a high fibre diet. The greatest feeding activity for horses found by Weinert *et al.* (2020) was in the late afternoon/early evening.

6.3 Future trends in horses monitoring

6.3.1 Surface EMG

Telemetric surface EMG is an emerging technology that has the potential to aid understanding of horse muscular function and improve performance (Williams *et al.*, 2013). Surface EMG offers a non-invasive tool that can quantify muscle activity onset and off-set of contractions, examine relationships between defined events and muscle performance and monitor muscle workload and adaptation over time (Hug *et al.*, 2010). However, as with other recording devices relying on sensor placement, problems with the contact may occur due to the presence of artefacts such as dirt, hair and irregular skin surface due to unspecified contaminants, all of which contribute to noise in the resulting data and potentially impaired reliability of data measurement.

6.3.2 Smart textiles

There is significant potential for so-called 'smart textiles' to contribute fully to the humane horse interface (McGreevy *et al.*, 2014). Depending on its behaviour, a smart textile system can be referred to as passive smart textiles (only a sensing function); active smart textiles (sensing a stimulus from the environment and react in it); very smart textiles (can adapt their behaviour to circumstances) (Tao, 2001). Smart textiles have been implemented in several ways to monitor the health of humans (Cherenack and Van Pieterson, 2012; Lymberis and Olsson, 2003; Pantelopoulos and Bourbakis, 2010). Integrating smart textiles with standard horse equipment will advance measurement taking and scientific research without distressing the horse or causing it pain. Given that horses are often highly reactive to external stimuli (Visser *et al.*, 2001) using smart textiles promises to deliver more reliable results than has been able to report so far because the measurement equipment can be integrated into gear that is familiar to horses.

6.3.3 Stable environment

Modern management of horses often requires them to be housed indoors for much of, if not all of the year. As a result, indoor light quality and levels can have a huge impact on their health, development and performance (Anonymous, 2018). Horses are conditioned to enter the fertile season when the days become longer because it has been conditioned to expect the availability of unlimited food supplies that can be used for growth. For horse breeders, it is important to understand the value of optimising the quantity and quality of lighting since we more and more require the stallion and mares and lactate outside of their natural breeding season. With lighting systems, the alteration of the day/night cycle can be calculated and accurately set so that a horse will naturally advance to the long day season quicker and also offset the start of the autumn. Sensor technologies creating natural dawn and dusk are becoming increasingly popular by horse breeders (Anonymous, 2018).

Traditional smoke detection technologies are problematic and unreliable in horse barns. Dust, moisture, ammonia in the air and bugs can clog up detection heads very quickly and false alarms can be very disruptive. Currently, optical smoke detectors are designed to work in stables and have been proven to be very effective (Anonymous, 2018).

6.3.4 Applications

This generation's fast-moving lives demands to have quick access to information and instant feedback on horses responses, behaviour, health and performance. Most wearables also provide accompanying apps to monitor the measurements. Some apps are strictly paired with a certain sensor, other apps can integrate and receive data from different sensors or companies. Also, the number of apps not connected to a sensor is growing rapidly. This category includes apps in which you can keep records of your horse's performance, rides and training, your horse's weight, your horse's health records, or evaluate the horse's experiencing pain.

References

Anonymous, 2018. Global Horsetech Market Report 14. Available at: https://tinyurl.com/47hsmwzk

Arkell, M., Archer, R.M., Guitian, F.J., May, S.A., 2006. Evidence of bias affecting the interpretation of the results of local anaesthetic nerve blocks when assessing lameness in horses. Veterinary Record 159: 346-349. https://doi.org/10.1136/vr.159.11.346

Arruda, T.Z., Brass, K.E., De La Corte, F.D., 2011. Thermographic assessment of saddles used on jumping horses. Journal of Equine Veterinary Science 31: 625-629. https://doi.org/10.1016/j.jevs.2011.05.011

Auclair-Ronzaud, J., Jousset, T., Dubois, C., Wimel, L., Jaffrézic, F., Chavatte-Palmer, P., 2020. No-contact microchip measurements of body temperature and behavioural changes prior to foaling. Theriogenology 157: 399-406. https://doi.org/10.1016/j.theriogenology.2020.08.004

Bartlett, E., Cameron, L.J., Marlin, D., 2018. Validation of the Orscana sensor to monitor equine thermal comfort. In: McDonnell, S., Padalino, B., Baragli, P. (eds.) 14th International Equitation Science Conference, Rome, Italy. Pisa Univeristy Press, Pisa, Italy, pp. 59.

Bertone, J.J., 2006. Excessive drowsiness secondary to recumbent sleep deprivation in two horses. Veterinary Clinics of North America – Equine Practice 22: 157-162. https://doi.org/10.1016/j.cveq.2005.12.020

Best, R., Standing, R., 2019. Feasibility of a global positioning system to assess the spatiotemporal characteristics of polo performance. Journal of Equine Veterinary Science 79: 59-62. https://doi.org/10.1016/j.jevs.2019.05.018

Broux, B., De Clercq, D., Vera, L., Ven, S., Deprez, P., Decloedt, A., Van Loon, G., 2018. Can heart rate variability parameters derived by a heart rate monitor differentiate between atrial fibrillation and sinus rhythm? BMC Veterinary Research 14: 7. https://doi.org/10.1186/s12917-018-1650-6

Casella, E., Khamesi, A.R., Silvestri, S., 2020. A framework for the recognition of horse gaits through wearable devices. Pervasive and Mobile Computing 67: 101213. https://doi.org/10.1016/j.pmcj.2020.101213

Cherenack, K., Van Pieterson, L., 2012. Smart textiles: challenges and opportunities. Journal of Applied Physics 112: 091301. https://doi.org/10.1063/1.4742728

Christensen, J.W., Munk, R., Hawson, L., Palme, R., Larsen, T., Egenvall, A., König von Borstel, U.U., Rørvang, M.V., 2021. Rider effects on horses' conflict behaviour, rein tension, physiological measures and rideability scores. Applied Animal Behaviour Science 234: 105184. https://doi.org/10.1016/j.applanim.2020.105184

Clayton, H.M., O'Connor, K.A., Kaiser, L.J., 2014. Force and pressure distribution beneath a conventional dressage saddle and a treeless dressage saddle with panels. Veterinary Journal 199: 44-48. https://doi.org/10.1016/j.tvjl.2013.09.066

Clayton, H.M., Singleton, W.H., Lanovaz, J.L., Cloud, G.L., 2005. Strain gauge measurement of rein tension during riding: a pilot study. Equine Comparative Exercise Physiology 2: 203-205.

Dong, J.G., 2016. The role of heart rate variability in sports physiology. Experimental and Therapeutic Medicine 11: 1531-1536. https://doi.org/10.3892/etm.2016.3104

DuBois, C., Zakrajsek, E., Haley, D.B., Merkies, K., 2015. Validation of triaxial accelerometers to measure the lying behaviour of adult domestic horses. Animal 9: 110-114. https://doi.org/10.1017/s175173111400247x

Duncan, P., 1980. Time-budgets of camargue horses. 2. Time-budgets of adult horses and weaned sub-adults. Behaviour 72: 26-49.

Ellis, A.D., Visser, E.K., Van Reenen, C.G., 2006. Effect of a high concentrate versus high fibre diet on behaviour and welfare of horses. In: 40th International Conference ISAE, Cranfield University, UK. University of Bristol, pp. 42.

Fenner, K., Yoon, S., White, P., Starling, M., McGreevy, P., 2016. The effect of noseband tightening on horses' behaviour, eye temperature, and cardiac responses. PLoS ONE 11: e0154179. https://doi.org/10.1371/journal.pone.0154179

Freire, R., Buckley, P., Cooper, J.J., 2009. Effects of different forms of exercise on post inhibitory rebound and unwanted behaviour in stabled horses. Equine Veterinary Journal 41: 487-492. https://doi.org/10.2746/095777309x383883

Frick, L., Schwarzwald, C.C., Mitchell, K.J., 2019. The use of heart rate variability analysis to detect arrhythmias in horses undergoing a standard treadmill exercise test. Journal of Veterinary Internal Medicine 33: 212-224. https://doi.org/10.1111/jvim.15358

Fries, M., Montavon, S., Spadavecchia, C., Levionnois, O.L., 2017. Evaluation of a wireless activity monitoring system to quantify locomotor activity in horses in experimental settings. Equine Veterinary Journal 49: 225-231. https://doi.org/10.1111/evj.12568

Fruehwirth, B., Peham, C., Scheidl, M., Schobesberger, H., 2004. Evaluation of pressure distribution under an English saddle at walk, trot and canter. Equine Veterinary Journal 36: 754-757. https://doi.org/10.2746/0425164044848235

Graubner, C., Gerber, V., Doherr, M., Spadavecchia, C., 2011. Clinical application and reliability of a post abdominal surgery pain assessment scale (PASPAS) in horses. Veterinary Journal 188: 178-183. https://doi.org/10.1016/j.tvjl.2010.04.029

Green, A., 2020. Measurement of horse core body temperature, University of Kentucky, Lexington, USA, 21 pp.

Green, A.R., Gates, R.S., Lawrence, L.M., 2005. Measurement of horse core body temperature. Journal Of Thermal Biology 30: 370-377. https://doi.org/10.1016/j.jtherbio.2005.03.003

Hale, L.A., Huggins, S.E., 1980. The electroencephalogram of the normal 'grade' pony in sleep and wakefulness. Comparative Biochemistry and Physiology – Part A 66: 251-257. https://doi.org/10.1016/0300-9629(80)90159-0

Hampson, B.A., Morton, J.M., Mills, P.C., Trotter, M.G., Lamb, D.W., Pollitt, C.C., 2010. Monitoring distances travelled by horses using GPS tracking collars. Australian Veterinary Journal 88: 176-181. https://doi.org/10.1111/j.1751-0813.2010.00564.x

HAS Hogeschool, 2016. Hippische monitor: perspectieven op de toekomst. HAS Hogeschool, Den Bosch, the Netherlands.

HAS Hogeschool, 2019. Hippische monitor: draagkracht en draagvlak. HAS Hogeschool, Den Bosch, the Netherlands.

Hebenbrock, M., Düe, M., Holzhausen, H., Sass, A., Stadler, P., Ellendorff, F., 2005. A new tool to monitor training and performance of sport horses using global positioning system (GPS) with integrated GSM capabilities. Deutsche Tierarztliche Wochenschrift 112: 262-265.

Hug, F., Turpin, N.A., Guével, A., Dorel, S., 2010. Is interindividual variability of EMG patterns in trained cyclists related to different muscle synergies? Journal of Applied Physiology 108: 1727-1736. https://doi.org/10.1152/japplphysiol.01305.2009

Ille, N., Erber, R., Aurich, C., Aurich, J., 2014. Comparison of heart rate and heart rate variability obtained by heart rate monitors and simultaneously recorded electrocardiogram signals in nonexercising horses. Journal of Veterinary Behaviour-Clinical Applications and Research 9: 341-346. https://doi.org/10.1016/j.jveb.2014.07.006

Kang, H., Zsoldos, R.R., Woldeyohannes, S.M., Gaughan, J.B., Guitart, A.S., 2020. The use of percutaneous thermal sensing microchips for body temperature measurements in horses prior to, during and after treadmill exercise. Animals 10: 1-20. https://doi.org/10.3390/ani10122274

KNHS, 2018. Nederland paardenland. KNHS, Ermelo, the Netherlands.

Krysiak, S., 2017. Horsemanship competence; study commissioned by the FEI. FEI, Lausanne, Switzerland. Available at: https://tinyurl.com/bzyk3f8h

Kurvers, C.M.H.C., Van Weeren, P.R., Rogers, C.W., Van Dierendonck, M.C., 2006. Quantification of spontaneous locomotion activity in foals kept in pastures under various management conditions. American Journal of Veterinary Research 67: 1212-1217. https://doi.org/10.2460/ajvr.67.7.1212

Lashley, M.J.J.O., Nauwelaerts, S., Vernooij, J.C.M., Back, W., Clayton, H.M., 2014. Comparison of the head and neck position of elite dressage horses during top-level competitions in 1992 versus 2008. Veterinary Journal 202: 462-465. https://doi.org/10.1016/j.tvjl.2014.08.028

Loomans, J.B.A., Stolk, P.W.T., Van Weeren, P.R., Vaarkamp, H., Barneveld, A., 2007. A survey of the workload and clinical skills in current equine practices in the Netherlands. Equine Veterinary Education 19: 162-168. https://doi.org/10.2746/095777307x186875

Lymberis, A., Olsson, S., 2003. Intelligent biomedical clothing for personal health and disease management: state of the art and future vision. Telemedicine Journal and e-Health 9: 379-386.

Marlin, D.J., Roberts, C.A., 1998. Qualitative and quantitative assessment of respiratory airflow and pattern of breathing in exercising horses. Equine Veterinary Education 10: 178-186. https://doi.org/10.1111/j.2042-3292.1998.tb00874.x

McGreevy, P., Winther-Christensen, J., König von Borstel, U., McLean, A., 2018. Equitation science. John Wiley and Sons Ltd., Hoboken, NY, USA.

McGreevy, P.D., Sundin, M., Karlsteen, M., Berglin, L., Ternstrom, J., Hawson, L., Richardsson, H., McLean, A.N., 2014. Problems at the human-horse interface and prospects for smart textile solutions. Journal of Veterinary Behaviour – Clinical Applications and Research 9: 34-42. https://doi.org/10.1016/j.jveb.2013.08.005

Morrison, R., Sutton, D.G.M., Ramsoy, C., Hunter-Blair, N., Carnwath, J., Horsfield, E., Yam, P.S., 2015. Validity and practical utility of accelerometry for the measurement of in-hand physical activity in horses. BMC Veterinary Research 11: 8. https://doi.org/10.1186/s12917-015-0550-2

Murase, H., Matsui, A., Endo, Y., Sato, F., Hada, T., 2018. Changes of lying behaviour in thoroughbred foals influenced by age, pasturing time, and weather conditions. Journal of Equine Science 29: 61-66. https://doi.org/10.1294/jes.29.61

Pantelopoulos, A., Bourbakis, N.G., 2010. A survey on wearable sensor-based systems for health monitoring and prognosis. IEEE Transactions on Systems, Man, and Cybernetics, Part C 40: 1-12. https://doi.org/10.1109/TSMCC.2009.2032660

Parker, M., Goodwin, D., Eager, R.A., Redhead, E.S., Marlin, D.J., 2010. Comparison of Polar® heart rate interval data with simultaneously recorded ECG signals in horses. Comparative Exercise Physiology 6: 137-142. https://doi.org/10.1017/S1755254010000024

Randle, H., Abbey, A., Sears, K., 2013. Qualitative (perceived) versus quantitative (actual) assessment of rein tension: what lessons can be learnt? In: Heleski, C.R., Wickens, C.L. (eds.) 9th International Equitation Science Conference, Delaware, USA, pp. 37.

Randle, H., Steenbergen, M., Roberts, K., Hemmings, A., 2017. The use of the technology in equitation science: a panacea or abductive science? Applied Animal Behaviour Science 190: 57-73. https://doi.org/10.1016/j.applanim.2017.02.017

Rietman, T.R., Stauffacher, M., Bernasconi, P., Aauer, J.A., Weishaupt, M.A., 2004. The association between heart rate, heart rate variability, endocrine and behavioural pain measures in horses suffering from laminitis. Journal of Veterinary Medicine 51: 218-225.

Rizzo, M., Arfuso, F., Giudice, E., Abbate, F., Longo, F., Piccione, G., 2017. Core and surface temperature modification during road transport and physical exercise in horse after acupuncture needle stimulation. Journal of Equine Veterinary Science 55: 84-89. https://doi.org/10.1016/j.jevs.2017.03.224

Robinson, T.R., Hussey, S.B., Hill, A.E., Heckendorf, C.C., Stricklin, J.B., Traub-Dargatz, J.L., 2008. Comparison of temperature readings from a percutaneous thermal sensing microchip with temperature readings from a digital rectal thermometer in equids. Journal of the American Veterinary Medical Association 233: 613-617. https://doi.org/10.2460/javma.233.4.613

Schmidt, A., Aurich, J., Möstl, E., Müller, J., Aurich, C., 2010. Changes in cortisol release and heart rate and heart rate variability during the initial training of 3-year-old sport horses. Hormones and Behaviour 58: 628-636. https://doi.org/10.1016/j.yhbeh.2010.06.011

Tao, X.M. (ed.), 2001. Smart fibres, fabrics and clothing. Fundamentals and applications. Woodhead Publishing, Cambridge, UK, 336 pp.

Ter Woort, F., Dubois, G., Didier, M., Van Erck-Westergren, E., 2021. Validation of an equine fitness tracker: heart rate and heart rate variability. Comparative Exercise Physiology 17: 189-198. https://doi.org/10.3920/CEP200028

Vendrig, C.C.C., 2013. Heart rate variability in endurance horses. How is the HRV affected by rest and different training settings? Utrecht University, Utrecht, the Netherlands.

Verdegaal, E., Delesalle, C., Caraguel, C.G.B., Folwell, L.E., McWhorter, T.J., Howarth, G.S., Franklin, S.H., 2017. Evaluation of a telemetric gastrointestinal pill for continuous monitoring of gastrointestinal temperature in horses at rest and during exercise. American Journal of Veterinary Research 78: 778-784. https://doi.org/10.2460/ajvr.78.7.778

Visser, E.K., Van Reenen, C.G., Hopster, H., Schilder, M.B.H., Knaap, J.H., Barneveld, A., Blokhuis, H.J., 2001. Quantifying aspects of young horses' temperament: consistency of behavioural variables. Applied Animal Behaviour Science 74: 241-258.

Visser, E.K., Van Reenen, C.G., Van der Werf, J.T.N., Schilder, M.B.H., Knaap, J.H., Barneveld, A., Blokhuis, H.J., 2002. Heart rate and heart rate variability during a novel object test and a handling test in young horses. Physiology and Behaviour 76: 289-296.

Visser, E.K., Van Wijk-Jansen, E.E.C., 2012. Diversity in horse enthusiasts with respect to horse welfare: an explorative study. Journal of Veterinary Behaviour – Clinical Applications and Research 7: 295-304. https://doi.org/10.1016/j.jveb.2011.10.007

Von Borell, E., Langbein, J., Després, G., Hansen, S., Leterrier, C., Marchant-Forde, J., Marchant-Forde, R., Minero, M., Mohr, E., Prunier, A., Valance, D., Veissier, I., 2007. Heart rate variability as a measure of autonomic regulation of cardiac activity for assessing stress and welfare in farm animals – a review. Physiology and Behaviour 92: 293-316. https://doi.org/10.1016/j.physbeh.2007.01.007

Waran, N., 2002. The welfare of horses. Kluwer Academic Publishers, Dordrecht, the Netherlands, 240 pp.

Waran, N., McGreevy, P., Casey, R.A., 2002. Training methods and horse welfare. In: Waren, N. (ed.) The welfare of horses. Kluwer Academic, Dordrecht, the Netherlands, pp. 151-180.

Weinert, J.R., Werner, J., Williams, C.A., 2020. Validation and implementation of an automated chew sensor-based remote monitoring device as tool for equine grazing research. Journal of Equine Veterinary Science 88: 11. https://doi.org/10.1016/j.jevs.2020.102971

Werner, J., Umstatter, C., Zehner, N., Niederhauser, J.J., Schick, M., 2016. Validation of a sensor-based automatic measurement system for monitoring chewing activity in horses. Livestock Science 186: 53-58. https://doi.org/10.1016/j.livsci.2015.07.019

Williams, J., Gundry, P., Richards, J., Protheroe, L., 2013. A preliminary evaluation of surface electromyography as a tool to measure muscle fatigue in the National Hunt racehorse. The Veterinary Nurse 4: 566-572. https://doi.org/10.12968/vetn.2013.4.9.566

World Horse Welfare, 2015. Removing the blinkers: the health and welfare of European Equidea in 2015. Available at: https://storage.googleapis.com/stateless-whwwp-screenbeetle-c/2019/09/b0d4fbeb-removing-the-blinkers-report.pdf.

Yarnell, K., Hall, C., Billett, E., 2013. An assessment of the aversive nature of an animal management procedure (clipping) using behavioural and physiological measures. Physiology and Behaviour 118: 32-39. https://doi.org/10.1016/j.physbeh.2013.05.013

Young, S., Stämpfli, H., Geor, R., McCutcheon, L.J., Pringle, J., 1995. Simple airflow direction and footfall sensors for use in equine exercise studies. Equine Veterinary Journal 27: 175-177. https://doi.org/10.1111/j.2042-3306.1995.tb04914.x

7. Sensors and automated monitoring in companion animals

M. de Kort[1*], K. Visser[2] and E. van Erp-van der Kooij[1]

[1]HAS University of Applied Sciences, P.O. Box 90108, 5200 MA 's Hertogenbosch, the Netherlands; m.dkort@has.nl
[2]Aeres University of Applied Sciences, P.O. Box 374, Dronten, the Netherlands

Highlights

- About 25% of European households have at least one dog or cat.
- In many families, the dog or cat is considered part of the family, and pet owners are willing to spend considerable money on their pets.
- The pet industry has grown large, with products such as (specialised) feed, clothing, toys and gadgets, and services such as animal physiotherapists and doggy day-care centres.
- Dog and cat owners are interested in their pet's health and behaviour. This interest is growing and stems from the role of the pet in the family as well as from the general interest in (health) monitoring of consumers: 'life-logging'.
- Trends in the pet sector are lifestyle and health.
- Concerns in the pet sector are individual housing of dogs and group housing of cats; keeping dogs and cats indoors; the consequences of breeding for exterior characteristics, and the welfare of therapy or assistance dogs.
- Monitoring the health and behaviour of dogs and cats can be done using many commercial devices, ranging from cameras in the house to wearable devices such as neck collars with sensors.
- An increasing amount of devices are focused on the communication between pet and owner, for example with cameras and remote control food dispensers.

7.1 Background of the companion animal sector

Dogs and cats play an important role in the household and family. It is often said there is no friend more loyal than a dog. Looking back at the development of the bond between humans and dogs or humans and cats, it appears that mutual interests have played a role. Living with humans provided the dog or cat with shelter from the elements, access to food and protection. For the human, the dog or cat provided company and in the case of dogs also protection and help during hunting. Ultimately, the relationship between humans and dogs has been further developed and we speak of an intensive domestication process in dogs, which today has led to a large number of very diverse dog breeds and a wide variety of tasks that dogs perform for humans. From assistance dogs helping people with a disability to dogs that assist people in the war, detection dogs, police dogs and rescue dogs. In cats, this domestication process has taken place much less intensively. There are increasing numbers of (owned) pets, with increased binding and interaction with owners. This leads to an increased economic value of the pet sector, but also to an increasing interest of owners to monitor pet health and behaviour.

7.1.1 The pet sector: numbers and economic value

The number of dogs in the world is estimated at 900 million. A significant proportion of these lives as stray dogs, estimated at 15-25%. The number of cats in the world is estimated at 600 million, of which a significant number also lives as strays. Dogs and cats that are kept as pets depend on people providing care, including food, drink and a place to rest (Van Heijst *et al.*, 2015). In the USA, there are more than 77 million dogs and 85 million cats; >60% of the US households have pets (Okin, 2017). Approximately 106 million cats and 87 million dogs are kept in Europe. In 2019, 85 million European households had at least one pet, with 25% of European households owning at least one cat and 24% of European households at least one dog. Within Europe, Russia has the largest population of dogs with 16 million dogs, followed by Germany with 10 million dogs. Russia also has the largest number of cats in Europe, almost 23 million, followed by Germany with 14 million cats. According to the NVG, the Netherlands has 1,950,000 dogs and 3,100,000 cats in 2020, which can be seen in Figure 7.1.

Figure 7.2 shows that 19.9% of households in the Netherlands have at least one dog and 24.6% at least one cat (Nederlandse Voedingsindustrie Gezelschapsdieren, 2020).

7. Sensors and automated monitoring in companion animals

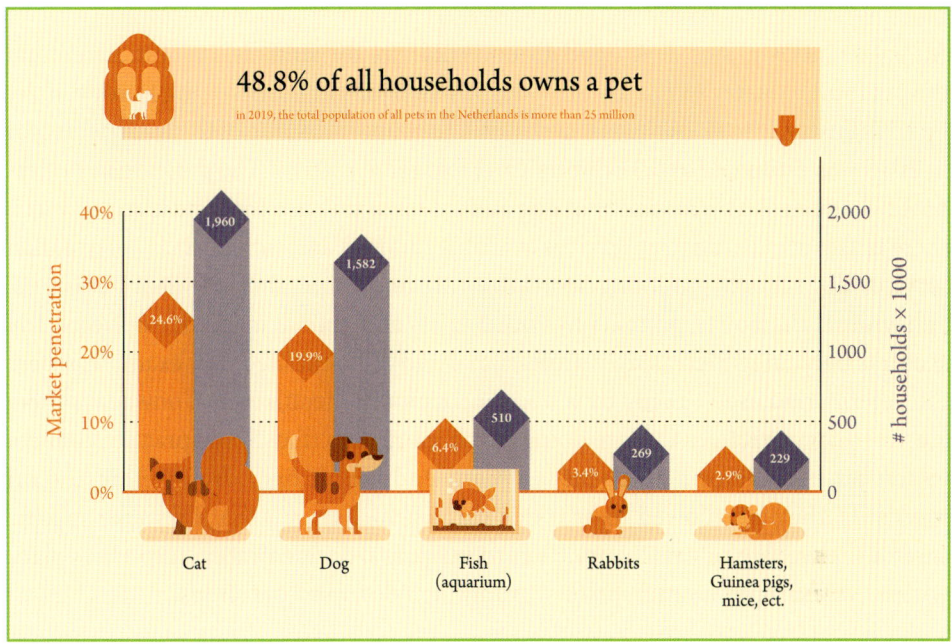

Figure 7.1. Ownership of dogs and cats in the Netherlands in 2020 (illustration: R. Schuijt).

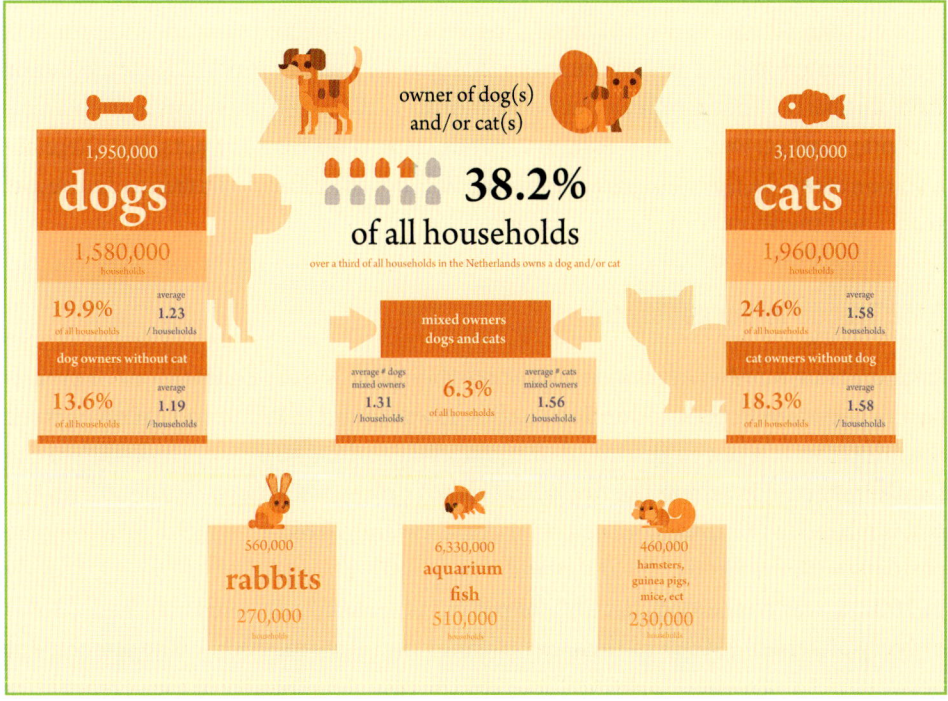

Figure 7.2. Pet ownership of Dutch households, 2020 (illustration: R. Schuijt).

An entire industry has developed around this care for animals. It is estimated that the pet food industry indirectly provides 900,000 jobs in Europe (The European Pet Food Industry, 2019). In the Netherlands, the economic value of the total pet sector in 2015 was estimated at € 3 billion (Van Heijst *et al.*, 2015). The pet industry offers products, such as a huge variety of dog and cat feeds but also toys and care products; there is also a wide range of services available for dogs and cats, such as veterinarian services, dog schools, grooming salons, day-care centres, pet shops, animal crematoria, behavioural therapists, animal physiotherapists and animal insurance options. The sector is becoming increasingly professional and differentiated. For example, where the veterinarian used to only treat dogs and cats with health issues, nowadays there are veterinary practices that are doing much more than that. Some practices adhere to the cat-friendly principle (https://catfriendlyclinic.org). By meeting certain standards, they are allowed to bear this name, which means that they take specific care of the welfare of cats. In addition to vet parks, differentiation also takes place within, for example, animal shelters, grooming salons and pet shops.

Doing business with dogs and cats appears to be an interesting business. In particular, the business dealing with pet supplies and animal feeds realises a high turnover. In 2015, this total turnover for all companion animals in the Netherlands was estimated at more than 1 billion euros. Globally, the value of pet-related services and products was estimated at 19.7 billion euros in 2019, with the pet food industry growing 2.6% annually (The European Pet Food Industry, 2019). Because this is such a fast-growing market, we see trends in new products and services for pets, but there are also concerns. In the next paragraphs, we will describe the main trends in the pet sector and mention some of the concerns.

7.1.2 Trends in the pet sector

An increasing number of owners appreciate their pets as close companions or even as part of their family. From adoption studies, it was found that dog and cat owners see the adoption of a pet as an ethical or moral obligation. The pet's relationship to the family, reflected in where the dog or cat sleeps or how much time is spent with the pet, is related to the amount of veterinary care the animal receives and to how long the pet stays with the family (Neidhart and Boyd, 2002; Sinski, 2016). From a US study in 2002, it was concluded that 86% of pet owners were 'somewhat attached' (36%) or 'very attached' (50%) to their pet (Clancy and Rowan, 2003). The attachment between pet and owner can be predicted by the complementarity of the personalities of pet and owner: the fit between the owner needs and the pet personality predicts the companion animal attachment (Bauer and Woodward, 2007). Dog and cat owners who are very attached to their pets are of great value to the pet industry, since they like to spoil their pet and spend a lot of money on the pet (American Pet Products Organisation, 2021).

In an overview of trends in the pet sector, we see that many trends are focused on lifestyle, matching owners' taste and pet accessories. About one-third of trends in the top 100 of pet trends (https://www.trendhunter.com/slideshow/2019-pet-trends) are gadgets or electronic devices for pets, varying from automated drinking and feeding bowls to robots and automated toys. Some devices monitor the health, feeding or drinking of the pets, record activity or track their location.

Lifestyle

In the dog and cat sector, lifestyle is playing an increasingly important role, as a result of the anthropomorphisation of pets. Dog and cat owners want the items they buy for the animals to fit into their home and lifestyle. Pet cushions come in various shapes, colours and sizes matching the interior, there is fashionable pet furniture available such as pet beds and food bowl holders, but also design enclosed litter boxes. There is an abundance of pet clothing, matching the owner's clothing, such as matching raincoats or pyjamas. There are hairbrushes for sale that give off a pleasant scent while brushing. It is even possible to design the whole apartment as cat-friendly or to build in toys and play material in the interior.

Health

Is being attached to the cat or dog also indicative of taking good care of the pet, and ensuring their health and welfare? There is a positive relationship between the pet's place in the household and veterinary care (Neidhart and Boyd, 2002). However, there is a thin line between taking good care of the pet and spoiling it. Overweight and obesity threaten the health of dogs and cats. In the US, 54% of dogs and 59% of cats are overweight or obese, causing premature death, pain and a high risk of other diseases, such as diabetes (German *et al.*, 2018). At the same time, we see an enormously growing interest in the specialised diets for dogs and cats, which is comparable to the growing interest in human nutrition and health. For example, the interest in vegetarian diets for dogs and cats is increasing with the interest in human vegetarian diets (Dodd *et al.*, 2019; Trendpanel Gezelschapsdieren, 2019).

A big trend in the health monitoring of pets are collars that monitor activity and behaviour. Monitoring feeding or drinking behaviour can be done with automatic feeding and drinking bowls; some devices use facial recognition to identify the pet and they monitor feeding or drinking behaviours. Several devices can be used to contact the pet when the owner is absent, via camera or sound systems and sometimes with the possibility to give a food reward via a remote access device. To substitute for the owner, there are interactive toys and robots for cats and dogs; a moving feeder that encourages hunting behaviour or a moving robot that activates the pet and aims to decrease separation anxiety.

7.1.3 Concerns in the pet sector

Alone or in a group

Dogs need social contact since they are group animals. Being home alone for a longer period affects the dog's wellbeing (Rehn and Keeling, 2011), although dogs can be trained to cope with the situation. In response to this, we see a growing number of doggy day-care and walking services, but also companies that offer the possibility to take your dog with you to the office (Trendpanel Gezelschapsdieren, 2019). Dogs being kept alone can develop chronic stress resulting in behavioural problems. When dogs that were kept indoors and alone were confronted with other dogs, for example, they showed more aggressive behaviours towards other dogs compared to dogs that were housed more socially (Beerda *et al.*, 1999a,b). Cats are more solitary animals. A growing concern for cats is the trend to keep several cats together in one household. Cats are also increasingly kept in groups in animal shelters. For the cats, this does not always turn out to be an ideal form of living together. Research shows that cats living alone in a household experience less stress than cats living with several other cats in one household (Finka *et al.*, 2014). Cats interact with each other in multi-cat households, but they do not appear to develop distinct dominance hierarchies or conflict solving strategies like some other species do. Therefore, they may attempt to prevent antagonistic encounters with other cats by avoiding them or decreasing their activity. Unrelated cats housed together in groups appear to spend less time interacting with each other than related cats do. Cats without close affiliative relationships prefer to have separate food and water sources, litterboxes, and resting areas to avoid unwanted interactions and competition for resources (Herron and Buffington, 2010).

Indoors versus outdoors

Europe, most cats live indoors but have outdoor access, while in the US, 63% of domestic cats are kept entirely indoors. There is a growing trend towards keeping cats exclusively indoors, especially when owners are younger (26-35 years old) and living in city centres or urban areas. Owners of indoor-only cats were mainly concerned about traffic accidents (Foreman-Worsley et al., 2021). In a study with 277 owners of indoor cats, 61% of the owners reported at least one problem behaviour of their cats, such as aggressive behaviour towards the owner or periuria (peeing at inappropriate places), although all of them played with their cat and 78% provided the cat with toys, available all of the time (Strickler and Shull, 2014). For cats kept indoors, it is important to be able to express their natural behaviours, such as scratching, chewing and elimination. When the cat is prohibited to show these behaviours (by not providing the appropriate surroundings to do so), health and behavioural problems will develop (Herron and Buffington, 2010). There are plenty of ideas and initiatives from organisations to help develop a cat-friendly surrounding in the house which also challenges the cat to develop a variety of new behaviour patterns. There are companies that design cat-friendly apartments or even overhead playgrounds for cats with walkways, cat perches and scratching poles, that you can install in your house (https://www.demilked.com/cat-playground-room-goldtatze). Keeping dogs indoors with no access to an outdoor area results in chronic stress and decreases their welfare. Dogs that were kept in small indoor kennels showed more auto grooming, paw lifting, vocalising, more incidents of coprophagy (eating poop) and repetitive behaviour (Beerda et al., 1999a; Spangenberg et al., 2006).

Breeding

An important issue in safeguarding animal welfare is the trend to breed companion animals with a cuter or so-called paedomorphic appearance. A companion animal with the appearance of large eyes, small nose and large ears is attractive for humans because of the intrinsic motivation of humans to care for babies and juveniles. This leads to breeding goals and directions with a preference for certain external characteristics, such as the shape of the face, the nose, the back and the animal's size, which leads to more and more extremities in certain dog and cat breeds. These extremities can lead to serious health problems for the dogs and cats in question. For example, short-nosed or flat-faced dogs such as bulldogs have extreme difficulty in breathing and consequently a major welfare concern. Where breeding on appearance is prominent, breeding inevitably also selects for certain behaviours and temperamental traits. Many pets, especially dogs, exhibit undesirable behaviour. This may lead to dangerous situations for the owner and often leads to bringing more dogs to shelters. More attention is needed for behavioural assessment and breeding for desirable behaviour in pets (King et al., 2012; Raad voor Dierenaangelegenheden, 2020).

Therapy and assistance dogs

Dogs in particular, but in some situations also cats, are increasingly being used in the health care of people. Within health care, there are several roles that a dog can fulfil, such as that of an assistance dog or a therapy dog. Assistance dogs can be guide dogs, hearing dogs or service dogs, specifically trained to perform several tasks to alleviate the effects of an individual's disability. Therapy dogs work in situations where psychotherapy or coaching therapy is offered to the client in short sessions with the help of a dog. The welfare of therapy and assistance dogs has not been studied much, but there are some concerns already reported. These include physical demanding tasks such as opening doors and pulling a wheelchair, but also mental strains such as lack of routine, lack of time off, being overweight and rough handling by children (Bremhorst et al., 2018). The demand for dogs as service dogs and as therapy dogs is growing, so fast that in the Netherlands this has led to a view on the animals in which it was investigated how the welfare of the animals can be guaranteed during such sessions (Raad voor Dierenaangelegenheden, 2019).

7.2 Sensor applications in companion animals

Many sensors are used for companion animals, ranging from cameras in the household and automatic cat doors to sensor collars monitoring behaviour or health. In this paragraph, we will focus mainly on wearable devices monitoring the behaviour and health of pets.

The use of activity meters is flourishing in the human mobile health technology industry. In 2012, the number of health-related apps was estimated at 40,000, with a value of approximately $ 718 million (Laing et al., 2014; Semper et al., 2016). The forecast for 2021 is that 929 million wearable devices will be used worldwide (Vailshery, 2021). These gadgets provide motivation for a healthier lifestyle and more exercise, help limit risks and thus contribute to an individual approach to preventive health care. This trend is also called 'life-logging': people monitoring a wide range of their daily activities, using apps, sensors and data, to improve, for example, their health and productivity (Arboleda Carpio et al., 2016; Rawassizadeh et al., 2013). This humane orientation and use of technology to monitor health and exercise have also found their way into the pet sector. Owners want to observe the activities and behaviours of their pets, to improve the wellbeing of their dog or cat, and understand how they act in the absence of the owner (Jukan et al., 2017; Rawassizadeh et al., 2013). In 2010 already a patent was filed for a 'human-dog communication system' containing several sensors such as a proximity meter, tilt sensor, GPS, accelerometer, temperature sensor, respiration meter, pulse meter, skin humidity sensor and a sniffer and signals such as a loudspeaker and light, aimed at determining the 'state' of the animal (Dror and Toder, 2010). Various technologies (gadgets) are now available to measure and provide insight into the health, activity and other variables of dogs and cats. These sensors predominantly focus on identification and tracking, behaviour monitoring, safety and security and health monitoring.

Cat and dog owners are eager to communicate with their pets. Animal-Computer Interface is a new branch of computer science, that focuses on understanding and improving human-animal communication. Research in ACI is linked to animal welfare and it involves technology that is designed to increase human-animal interaction and that is based on the needs of the animal (Jukan et al., 2017; Mancini, 2011). Research in ACI is diverse, however, there is a focus on positive stimulation of the environments of the animals, for example by improving play behaviour in cats with digital games (Pons et al., 2014). In dogs, devices are developed that stimulate communication between dogs and humans, for example with a smartphone attached to the dog (Lemasson et al., 2013). For commercial purposes, devices are available like the Clever Pet Hub (https://clever.pet), where owners can communicate with their dog, play a game and provide the dog with a threat. The device can be controlled by their smartphone.

7.2.1 Physiological measures

Heart rate

Heart rate monitors are used in research in dogs and cats, and from there have found their way into the commercial market. In this section, studies in pets with heart rate monitors are summarised and heart rate sensors for the pet market are described.

With heart rate monitors a non-invasive measurement of the cardiovascular function can be performed under different field conditions. Heart rate (HR) can be used as a measure for activity and is linked with results from activity meters (Ortmeyer et al., 2018). HR measurements can be used to evaluate mental states, such as fear and anxiety or bonding with the owner, and animal health (Jackson et al., 2014; Kuhne et al., 2014; Ogata et al., 2006). Heart rate variability (HRV), the variability between subsequent heartbeats, is a measure of autonomic tone, influenced by psychophysiological factors, neurohormonal mechanisms and cardiac disease. HRV measures the response of the autonomic nervous system and especially the balance between the parasympathetic and sympathetic branch (Von Borell et al., 2007).

Heart rate measurements in practice

There have been studies into a wireless ECG sensor for dogs, with pointed style electrodes in the collar or wireless body electrodes. This is a good alternative for a standard ECG and can be used in a practical situation, with the dog lying down or walking around (Brugarolas et al., 2014; Krvavica et al., 2016). Heart rate can also be used as an indicator of fear, stress or emotions in dogs, where some say that heart rate is a more reliable parameter than behaviour (Fallani et al., 2007; Katayama et al., 2016; Ogata et al., 2006) and that with HRV it is possible to differentiate between positive and negative emotional states (Zupan et al., 2016); other studies show that behaviour (e.g. tail wagging) and eye temperature are better indicators of positive emotional state (Travain et al., 2016). In response to emotional or stressful situations, heart rate increases (Kuhne et al., 2014). This can be the result of stress or fear but also of (positive) anticipation. In a study with 17 privately owned dogs of different breeds, the animals showed the same increase in heart rate in reaction to the entrance of the owner as to the entrance of a stranger (Palestrini et al., 2005). In therapy dogs, it seems that behavioural response can be suppressed while heart rate responses do indicate an emotional reaction (Fallani et al., 2007; Palestrini et al., 2017).

Heart rate and HRV measurements in research

HR and HRV have been measured in dogs using several validated heart rate monitors, designed as a wearable sensor (Essner et al., 2013, 2015a,b; Jonckheer-Sheehy et al., 2012). The gold standard for HR and HRV is the electrocardiogram (ECG). Heart rate sensors usually record R-R intervals, which is the time between successive heartbeats and HRV is recorded as the variation of the R-R intervals. The root mean square of successive differences of inter-beat intervals (RMSSD) reflects the beat-to-beat variance in HR. Often also the power in the high frequency (HF) or low frequency (LF) range is recorded, which might be associated with activity of the sympathetic and parasympathetic nervous system, which reacts to stress.

HRV can be used to detect possible signs of heart disease such as cardiomyopathy (Minors and O'Grady, 1997) but can also be used to assess mental state. It is believed that HRV can be used to indicate aggression and fear in dogs. In a small study with 10 dogs with anxiety-related problems and 20 controls, the problem dogs showed reduced HRV (Wormald et al., 2017). In another study involving 32 dogs, it was found that the 16 dogs with a history of bite incidents had a lower HRV than the dogs with no record of bite incidents; furthermore, the aggression reported by the owners was negatively related to HRV (Craig et al., 2017). Some HRV parameters are associated with a positive emotional state. In a Swedish study with nine beagles, the HR and HRV of the dogs were compared after getting a food reward (meatball or food pellet) or a social reward (interaction with a person). A decrease in HF was associated with the meatball stimulus compared to the food pellet; the dogs liked the meatball more than they liked the food pellet. The reward phase, where the dogs could interact with the person or eat the food, was associated with a decrease in HF and RMSSD compared to the preceding phase, where they could only look at the person or food. This suggests that parasympathetic deactivation is associated with a more positive emotional state in the dog (Zupan et al., 2016). Dogs that are exposed to classical music show higher HRV, which is considered positive (Bowman et al., 2015; Köster et al., 2019).

A specific device to monitor cardiac rhythm in humans and now also in dogs and cats is a Holter monitor. This monitor is a non-invasive tool for the monitoring of the cardiac rhythm over a prolonged period during normal daily activities. Electrodes are applied to the chest of the dog or cat and in most cases, the recorder is attached to the animals back. In small breed dogs and most cats, the size of the monitor may prohibit the animal from moving around and in these cases, the Holter monitor can be placed next to the animal in the cage for the desired recording period. In free-moving animals, wearing the Holter monitor can influence the daily activity pattern of the animal, as was shown in a study with 32 cats with and without arrhythmias. More than half of the owners reported that wearing the monitor affected the daily activity of their cat. By using this device it has been demonstrated that the incidence of serious arrhythmias is much greater than previously determined using routine electrocardiography. An electrocardiogram is typically recorded for a short period (seconds to minutes), and it is likely to miss or underestimate the underlying arrhythmia this way (Jackson et al., 2014; Petrie, 2005).

Temperature

In an Australian study with feral cats, the temperature was been studied using remote sensing. Temperature sensors were implanted in the abdomen of eight cats after trapping them and temperatures were recorded for 14 days. The cats had to be caught, to be operated on to implant the sensors, and in order to retrieve the sensors the cats were trapped with poisonous bait and the sensors were cut out of the cadavers. (Hilmer et al., 2010). This is not a very cat-friendly measuring method and fortunately, there are easier and non-invasive ways to automatically monitor the temperature. In cats and dogs, body temperature can be monitored via a sensor in the collar (Ortmeyer *et al.*, 2018).

Although the gold standard for body temperature is rectal temperature, thermal cameras using infrared can also be used to measure body temperature. In dogs, in one study a good relation was found between eye and ear temperature measured with an infrared thermometer and rectal temperature; ear temperature gave the best correlation with rectal temperature. All three measuring methods could detect hyperthermia in dogs (Zanghi, 2016). However, in another study, a poor correlation (r=0.4) was found between eye temperature measured with a non-contact infrared thermometer and rectal temperature (Kreissl and Neiger, 2015). Infrared cameras can also be used to detect diseases in dogs. In a veterinary study, it was shown that dogs with mammary tumours showed higher thermographic temperatures in the mammary glands than healthy dogs (Pavelski *et al.*, 2015). In cats, thermographic cameras can be used similarly, for example, to detect hyperthyroidism. In a study with 17 cats with hyperthyroidism and 12 control cats, thermal imaging of the neck could classify between ill and healthy cats. The accuracy was 81% (unshaved neck) and 88% (shaved neck) (Waddell *et al.*, 2015). Although some studies show relatively good correlations between rectal and thermographic temperatures, others do not, and practical experience shows that many factors can influence the reliability of results, such as environmental conditions, distance, hair colour and cleanliness of the fur.

Respiration rate

Respiration rate is an important health indicator. Respiration can be monitored in dogs using an implanted pressure sensor. However, this is an invasive method, aimed at health research (Brugarolas *et al.*, 2014). Respiration rate can also be monitored non-invasively using a combination of an accelerometer and a magnetic impedance sensor; combining these sensors gave good results for respiration in sitting dogs, but not in walking dogs (Antink *et al.*, 2019). Automatic respiration measurement can also be done via a sensor in the collar (Ortmeyer *et al.*, 2018). A new method is being developed to monitor respiration in dogs and cats using UWB radar. With this method, the respiration rate of cats and dogs can be recorded from a distance (Wang *et al.*, 2020).

Glucose

In cats or dogs with diabetes, measuring blood glucose level can be stressful because the footpads, ears or lips of the animal have to be pricked. With a flash monitoring system, glucose levels in the interstitial fluid (under the skin) are monitored using an implanted sensor. The device consists of a small round plate with a flexible sampling catheter. The sensor can be used for 14 days in a row and measures glucose levels continuously; the sensor has to be scanned regularly to capture the data. The device is reported to be easy to use and accurate (Corradini *et al.*, 2016; Hammond-Lenzer *et al.*, 2019).

7.2.2 Behavioural measures

Activity levels

For measuring activity normally accelerometers are used. These are non-invasive monitoring devices that record changes in acceleration that relate to the intensity, frequency, duration of movement and pattern of activity. Activity meters for pets have become increasingly popular. But as with all sensors, its validity must be studied. Several accelerometers have found to be valid for measuring activity in dogs and cats (Andrews *et al.*, 2015; Belda *et al.*, 2018; Ortmeyer *et al.*, 2018; Yamazaki *et al.*, 2020). In attaching an accelerometer to a dog collar, it is important that when walking the dog, the leash is not attached to the same collar as the accelerometer and that it is not loosely zip-tied to the collar, since this gives inaccurate readings (Martin *et al.*, 2017). For dogs, it was found that there is no difference in activity count for larger or smaller dogs with longer or shorter limbs during controlled activity (Thompson, 2018). Since activity levels for cats can vary considerably between individuals, each animal should be considered independently and when assessing changes in activity levels, each animal should serve as its control (Andrews *et al.*, 2015).

Accelerometers can be used to measure activity level in behavioural studies (Patel *et al.*, 2017) (Figure 7.3) and in veterinary medicine to monitor the progression of disease or efficacy of a treatment (Belda *et al.*, 2018). In dogs with osteoarthritis, a common form of arthritis causing pain of the joints and lameness, it was shown that the activity of treated dogs, measured with an accelerometer, improved compared to the group that received a placebo (Brown *et al.*, 2010a). Activity meters have been used to compare dogs with pruritis to healthy dogs: this skin disease caused scratching, which can be detected with an accelerometer (Nuttall and McEwan, 2006). Accelerometers can be used to study cat activity in for example diet studies or clinical trials, where activity is one of the factors to determine the response to the diet, feeding method or the medication (Coleman *et al.*, 2020; Naik *et al.*, 2018; Thomas *et al.*, 2017).

Figure 7.3. Dog with an accelerometer (Fitbark) (photo: Julia van Knippenberg).

Older and larger or overweight dogs tended to show lower activity counts than younger and smaller dogs if given freedom (Brown *et al.*, 2010b). Substantial weight loss was not associated with a spontaneous increase in physical activity or with a reduction in sedentary behaviour of dogs (Morrison *et al.*, 2014). In a study with 60 cats at the Hills Pet Nutrition Center, activity data were gathered using collar-mounted triaxial accelerometers. No activity, low activity, medium activity and high activity was detected using thresholds for the dataset. During the two weeks of observation, only three of the cats removed the sensors. It was found that highly active cats could be detected from the activity meter data, but it was not possible to distinguish between medium and low activity levels. Older cats were less active than younger cats, even excluding cats <1 year old, and although the overweight cats seemed to move less, no difference in the data could be found (Thompson, 2018).

Specific behaviours and emotions

From accelerometer data, specific behaviours can be determined. In dogs it is possible to detect active behaviours, such as walking, trotting, galloping, eating, head shaking and sniffing from accelerometer results, but it is difficult to distinguish between the stationary behaviours standing, sitting and lying (Den Uijl *et al.*, 2017; Kumpulainen *et al.*, 2018). Also for cats, behaviour can be linked to accelerometer data. In a study with 31 privately owned cats, behaviours were scored in cats wearing activity meters. Similar to the dog study, the classifier for moving behaviours such as walking, running and jumping performed well, just as for behaviours with a specific posture such as lying, sitting or eating. Defecation and urination were difficult to detect because the movement and posture were similar to other behaviours. Location sensors could solve this problem since these behaviours usually occur at a specific place (Thompson, 2018). In a study with four laboratory cats, an accelerometer was attached to the head of the animals in combination with implanted brain electrodes for electromyograms (EMG) and electrooculograms (EOG). This way, movement and behaviour could be linked to brain activity (Grand *et al.*, 2013).

In a study with ten dogs wearing neck and tail sensors containing an accelerometer and a gyroscope, behaviours and positive, neutral or negative emotions were detected. Using labelled video, classification models were trained to detect positive, neutral and negative emotions from the tail movement, and the activities walking, sitting, 'stay,' eating, 'sideway,' jumping, and nose work from the neck movement. Various machine learning techniques were used. For both emotion and activity detection, very high precision and accuracy rates were achieved, with the artificial neural network performing best with an accuracy of 0.93 and a precision of 0.95 for emotion detection and an accuracy of 0.97 and a precision of 0.92 for activity detection (Aich *et al.*, 2019).

Location and use of space

For cats, that often can move about freely, there is an increasing interest from the owners to know their whereabouts (Figure 7.4). Outdoor, GPS can be used to determine location. In studies using GPS collars or radiotracking, home ranges of farm and pet cats were determined. It was concluded that male cats have larger home ranges than female cats and that the home range of pet cats from rural areas was much larger than that of pet cats from urban areas (Hall *et al.*, 2016). From studies with domestic cats wearing GPS collars, it was found that the daily area range was 1.94 ha. (Thomas *et al.*, 2014). Cat density of rural cats can be estimated using camera

Figure 7.4. Cat with a GPS tracker (photo: Manon de Kort).

traps (Hansen *et al.*, 2018). Location sensors for pet cats and dogs are used to find runaways or as a preventive measure against theft.

Indoors, Ultra-Wide Band tracking can be used to track animals and perform accurate behavioural studies. At the Royal Canin's Cattery in France, six cats living in a group were tracked for 14 days using UWB tracking collars of Ubisense. Seven antennas with a detection rate of 60 meters were used, resulting in an accuracy of 15 cm. Travelling distance, use of space and interaction between cats and with humans could be measured (Parker *et al.*, 2017). Indoor tracking systems can also be used to study activity patterns and daily rhythms (Parker *et al.*, 2019).

7.2.3 Sound

Barking in dogs can be detected with sound sensors or motion sensors. In search and rescue (SAR) dogs that were trained to find victim by smell and react with continuous barking, sensor vests were used. Based on audio, continuous barking could be detected with an accuracy of 0.95, while based on motion, the accuracy of detection was 0.9 (Komori et al., 2015). To differentiate between the vocalisations of dogs, convolutional network analysis can be used to analyse intensity data of sounds, recorded with a noise sensor. Using this type of analysis it is possible to detect barking, growling, howling and whining (Kim et al., 2018). In pet dogs, barking is often considered unwanted behaviour. Anti-bark collars have been designed with a vibration sensor in the collar that detects barking. When a dog barks, the collar gives a negative signal ('correction') to the dog, which can be a (lemon) spray, ultrasound or an electrical stimulus, to prevent the dog from further barking. Dogs wearing electronic or lemon spray collars bark less than control dogs, and in a study with 24 dogs, no differences were found between activity or stress levels of the dogs wearing electronic collars, lemon spray collars or inactivated collars (controls) (Groh and Lee, 2003; Steiss et al., 2007). The use of shock collars is banned in Denmark, Norway, Sweden, Austria, Switzerland, Slovenia, and Germany, and in some territories of Australia, including New South Wales and South Australia. Electronic dog collars are also banned in the Netherlands since 1 July 2020. In England, a ban has been announced already in 2018 but the legislation has been delayed until at least 2021. An alternative method to reduce barking in dogs automatically has been tested in a study with five pet dogs, that were left alone for 20 minutes in an experimental setting. Each dog would receive a food reward after a fixed period of not barking. Barking was detected automatically by sound analysis software, and the food reward was delivered with a remote feeder. This training reduced barking significantly (Protopopova et al., 2016).

In a study, it was shown that humans can correctly categorise dog barks into different situations, and also score the emotional content of the barks. In this study, humans could differentiate between aggressive and fearful barks based on the pitch of the sound (Pongrácz et al., 2006). This classification of dog barks into context, emotion or intensity can also be done automatically applying machine learning models (Hantke et al., 2018). Recently, a collar that claims to detect the emotional state of the dog from the barking sound has been brought to the market (www.petpuls.net). The device applies voice recognition models and its algorithm detects five emotional states from the barking sounds of the dog: happy, relaxed, anxious, angry or sad.

7.2.4 Summary of sensors used in companion animals

An overview of sensors used in cats and dogs is given in Table 7.1.

Table 7.1. Overview of sensor used in companion animals, in development or commercially available.

PLF sensor	Animal species	Where	What it measures	Why	Company website
Camera	cats and dogs	in house	behaviour	health, stress	Several manufacturers
		on-device	behaviour, stress, feeding, drinking	interaction with pet	Store.skymee.com
	dogs	back harness	behaviour, activity	health, stress	https://tinyurl.com/y47shwkg
Heart rate sensor	cats and dogs	breast (moveable)	heart rate	health	Stemoscope.com
	dogs	chest harness	heart rate	health, stress training	Polar.com
		neck collar	heart rate	health, stress	Petpace.com
				health	Waggit.dog
Temperature sensor	dogs	neck collar	skin temperature	health	Waggit.dog; Petpace.com
Accelerometer	cats and dogs	neck collar	activity, positions	health	Petpace.com
			activity	health	Petkit P2
	dogs	neck collar	activity, positions	health	Fitbark.com; Waggit.com
			activity	health	Whistle.com; www.theactigraph.com
		hanger on collar	activity	health	PetAlways; PitPat.com; Justpoochplay.com; KippyVita.eu/it
GPS	cats	hanger on collar	location	lost and theft	Mytabcat.com
	cats and dogs	hanger on collar	location	lost and theft	Tractive.com
	dogs	neck collar	location	lost and theft	Waggit.dog; Fitbark2; PetPace.com
		hanger on collar	location	lost and theft	KippyVita.eu/it
Respiration sensor	cats and dogs	neck collar	rate changes	health	Petpace.com
Weighing sensor	cats and dogs	bowl	weight of eaten food	health	Pet Always IBowl
Chemical sensor	dogs	under the skin	glucose	health	www.FreestyleLibre.us
Vibration sensor	dogs	neck collar	barking	training	Garmin.com; Petsafe.net; Aetertek.com

7.3 Future trends in technology for companion animals

Sensors will probably be much more common in the everyday life of cats and dogs. The US market for pet wearables is expected to grow from 703 million USD in 2019 to 1,718 million USD in 2024 (BusinessWire, 2019).

The use of GPS sensors for dogs and cats has been established and increasingly advanced activity meters for monitoring the movement and health of dogs and cats are coming up fast. The demand for more insight into pet health will increase, which will further develop the pet sensor market. Advanced developments are automatic feeding systems, recognising individual collar tags for use in multiple-cat households, that monitor feeding and drinking behaviour (https://catspad.com/automatic-cat-feeder) or automatic feeders that recognise individuals based on facial recognition (Stampler, 2014). With such a device, personalised food can be offered in a household with several cats (Conolly, 2017). Other examples of automated control by sensors are wireless pet fences connected to a shock collar, and dog bark collars.

Facial recognition and the recognition of vocalisations to detect stress and discomfort are subject of study. Furthermore, the possibility to use artificial intelligence to detect emotions and patterns of animal sounds is being studied (Overbay, 2018).

Another important development that would contribute to more insight and progress in the field of working with the results of sensor technology is more research and open-source data (Ladha, n.d.). At this moment research is being done into a sensor and platform that combines various technologies, the Voyce Health and Wellness Ecosystem. The platform offers possibilities for an Internet of Things solution in which various forms and applications of sensors are combined (www.voyce.com). An additional trend might be the integration of several sensor systems.

Many owners use GoPro-like cameras or surveillance cameras to monitor their pets. Both types of cameras are also used to interact at distance with the pets, often with help of apps. Another way of monitoring and interacting with pets are automated toys and food dispensers (or cat flaps), often with a camera or wireless connection to an app. Apps are increasingly used to automatically trigger specific events in response to particular behaviours or states.

References

Aich, S., Chakraborty, S., Sim, J.S., Jang, D.J., Kim, H.C., 2019. The design of an automated system for the analysis of the activity and emotional patterns of dogs with wearable sensors using machine learning. Applied Sciences 9: 4938. https://doi.org/10.3390/APP9224938

American Pet Products Organisation, 2021. Pet industry market size & ownership statistics. Available at: https://www.americanpetproducts.org/press_industrytrends.asp.

Andrews, C.J., Potter, M.A., Thomas, D.G., 2015. Quantification of activity in domestic cats (*Felis catus*) by accelerometry. Applied Animal Behaviour Science 173: 17-21. https://doi.org/10.1016/j.applanim.2015.05.006

Antink, C.H., Pirhonen, M., Väätäjä, H., Somppi, S., Törnqvist, H., Cardó, A. V, Teichmann, D., Vainio, O., Surakka, V., Vehkaoja, A., 2019. Sensor fusion for unobtrusive respiratory rate estimation in dogs. IEEE Sensors Journal 19: 7072-7081. https://doi.org/10.1109/JSEN.2019.2912002

Arboleda Carpio, S.L., Sohail, S., Clark, K., Fagan, J.M., 2016. Fitness gadgets as a form of preventative healthcare. Rutgers, The State University of New Jersey, NJ, USA. Available at: https://doi.org/10.7282/T38W3GN0

Bauer, A., Woodward, L., 2007. People and their pets: a relational perspective on interpersonal complementarity and attachment in companion animal owners. Society and Animals 15: 169-189. https://doi.org/10.1163/156853007X187117

Beerda, B., Schilder, M.B.H., Bernadina, W., Van Hooff, J.A.R.A.M., De Vries, H.W., Mol, J.A., 1999a. Chronic stress in dogs subjected to social and spatial restriction. II. Hormonal and immunological responses. Physiology and Behavior 66: 243-254. https://doi.org/10.1016/S0031-9384(98)00290-X

Beerda, B., Schilder, M.B.H., Van Hooff, J.A.R.A.M., De Vries, H.W., Mol, J.A., 1999b. Chronic stress in dogs subjected to social and spatial restriction. I. Behavioral responses. Physiology and Behavior 66: 233-242. https://doi.org/10.1016/S0031-9384(98)00289-3

Belda, B., Enomoto, M., Case, B.C., Lascelles, B.D.X., 2018. Initial evaluation of PetPace activity monitor. Veterinary Journal 237: 63-68. https://doi.org/10.1016/j.tvjl.2018.05.011

Bowman, A., Scottish, S.P.C.A., Dowell, F.J., Evans, N.P., 2015. 'Four Seasons' in an animal rescue centre; classical music reduces environmental stress in kennelled dogs. Physiology and Behavior 143: 70-82. https://doi.org/10.1016/j.physbeh.2015.02.035

Bremhorst, A., Mongillo, P., Howell, T., Marinelli, L., 2018. Spotlight on assistance dogs – legislation, welfare and research. Animals 8: 129. https://doi.org/10.3390/ani8080129

Brown, D.C., Boston, R.C., Farrar, J.T., 2010a. Use of an activity monitor to detect response to treatment in dogs with osteoarthritis. Journal of the American Veterinary Medical Association 237: 66-70. https://doi.org/10.2460/javma.237.1.66

Brown, D.C., Michel, K.E., Love, M., Dow, C., 2010b. Evaluation of the effect of signalment and body conformation on activity monitoring in companion dogs. American Journal of Veterinary Research 71: 322-325. https://doi.org/10.2460/ajvr.71.3.322

Brugarolas, R., Dieffenderfer, J., Walker, K., Wagner, A., Sherman, B., Roberts, D., Bozkurt, A., 2014. Wearable wireless biophotonic and biopotential sensors for canine health monitoring. Proceedings of IEEE Sensors 2014: 2203-2206. https://doi.org/10.1109/ICSENS.2014.6985477

BusinessWire, 2019. Pet wearable market by component (GPS Chips, RFID chips, sensors, Wi-Fi, cellular, Bluetooth chips, processors, memory, displays, batteries), product (smart collars, smart cameras, smart harnesses), and region – Global Forecast to 2024. Research and Markets, Dublin, Ireland. Available at: https://tinyurl.com/3k34fv6v.

Clancy, E., Rowan, A., 2003. Companion animal demographics in the United States: a historical perspective. The State of the Animals II, 19 pp. Available at: https://tinyurl.com/2waspah9.

Coleman, A.E., DeFrancesco, T.C., Griffiths, E.H., Lascelles, B.D.X., Kleisch, D.J., Atkins, C.E., Keene, B.W., 2020. Atenolol in cats with subclinical hypertrophic cardiomyopathy: a double-blind, placebo-controlled, randomized clinical trial of effect on quality of life, activity, and cardiac biomarkers. Journal of Veterinary Cardiology 30: 77-91. https://doi.org/10.1016/j.jvc.2020.06.002

Conolly, A., 2017. Futuristic fido: tech that's reinventing the pet world. Alltech, Lexington, KY, USA. Available at: https://tinyurl.com/39dnth26.

Corradini, S., Pilosio, B., Dondi, F., Linari, G., Testa, S., Brugnoli, F., Gianella, P., Pietra, M., Fracassi, F., 2016. Accuracy of a flash glucose monitoring system in diabetic dogs. Journal of Veterinary Internal Medicine 30: 983-988. https://doi.org/10.1111/jvim.14355

Craig, L., Meyers-Manor, J.E., Anders, K., Sütterlin, S., Miller, H., 2017. The relationship between heart rate variability and canine aggression. Applied Animal Behaviour Science 188: 59-67. https://doi.org/10.1016/j.applanim.2016.12.015

Den Uijl, I., Gómez Álvarez, C.B., Bartram, D., Dror, Y., Holland, R., Cook, A., 2017. External validation of a collar-mounted triaxial accelerometer for second-by-second monitoring of eight behavioural states in dogs. PLoS ONE 12: e0188481. https://doi.org/10.1371/journal.pone.0188481

Dodd, S.A.S., Cave, N.J., Adolphe, J.L., Shoveller, A.K., Verbrugghe, A., 2019. Plant-based (vegan) diets for pets: a survey of pet owner attitudes and feeding practices. PLoS ONE 14: e0210806. https://doi.org/10.1371/journal.pone.0210806

Dror, J.S., Toder, H., 2010. System and method for human dog communication. Patent US 2010/0231391. Available at: https://patents.google.com/patent/US6192835B1/en.

Essner, A., Sjöström, R., Ahlgren, E., Gustås, P., Edge-Hughes, L., Zetterberg, L., Hellström, K., 2015a. Comparison of Polar® RS800CX heart rate monitor and electrocardiogram for measuring inter-beat intervals in healthy dogs. Physiology and Behavior 138: 247-253. https://doi.org/10.1016/j.physbeh.2014.10.034

Essner, A., Sjöström, R., Ahlgren, E., Lindmark, B., 2013. Validity and reliability of Polar® RS800CX heart rate monitor, measuring heart rate in dogs during standing position and at trot on a treadmill. Physiology and Behavior 114-115: 1-5. https://doi.org/10.1016/j.physbeh.2013.03.002

Essner, A., Sjöström, R., Gustås, P., Edge-Hughes, L., Zetterberg, L., Hellström, K., 2015b. Validity and reliability properties of canine short-term heart rate variability measures-a pilot study. Journal of Veterinary Behavior – Clinical Applications and Research 10: 384-390. https://doi.org/10.1016/j.jveb.2015.05.006

Fallani, G., Prato Previde, E., Valsecchi, P., 2007. Behavioral and physiological responses of guide dogs to a situation of emotional distress. Physiology and Behavior 90: 648-655. https://doi.org/10.1016/j.physbeh.2006.12.001

Finka, L.R., Ellis, S.L.H., Stavisky, J., 2014. A critically appraised topic (CAT) to compare the effect of single and multi-cat housing on the physiological and behavoural measures of stress in domestic cats in confined environments. BMC Veterinacy Research 10: 73. https://doi.org/10.1186/1746-6148-10-73

Foreman-Worsley, R., Finka, L.R., Ward, S.J., Farnworth, M.J., 2021. Indoors or outdoors? An international exploration of owner demographics and decision making associated with lifestyle of pet cats. Animals 11: 1-25. https://doi.org/10.3390/ani11020253

German, A.J., Woods, G.R.T., Holden, S.L., Brennan, L., Burke, C., 2018. Dangerous trends in pet obesity. Veterinary Record 182: 25. https://doi.org/10.1136/vr.k2

Grand, L., Ftomov, S., Timofeev, I., 2013. Long-term synchronized electrophysiological and behavioral wireless monitoring of freely moving animals. Journal of Neuroscience Methods 212: 237-241. https://doi.org/10.1016/j.jneumeth.2012.10.008

Groh, W.S., Lee, A.L., 2003. Spray control anti-bark collar. Patent US 6668760B2. Available at: https://patents.google.com/patent/US6668760B2/en.

Hall, C.M., Bryant, K.A., Haskard, K., Major, T., Bruce, S., Calver, M.C., 2016. Factors determining the home ranges of pet cats: A meta-analysis. Biological Conservation 203: 313-320. https://doi.org/10.1016/j.biocon.2016.09.029

Hammond-Lenzer, M., Wilson, S., Wasik, B., 2019. Flash glucose monitoring with freestyle libre in cats and dogs. VeterinaryPartner. Available at: https://tinyurl.com/ms99w3pm.

Hansen, C.M., Paterson, A.M., Ross, J.G., Ogilvie, S.C., 2018. Estimating feral cat (*Felis catus*) density in a rural to urban gradient using camera trapping. New Zealand Journal of Zoology 45: 213-226. https://doi.org/10.1080/03014223.2018.1494609

Hantke, S., Cummins, N., Schuller, B., 2018. What is my dog trying to tell me? the automatic recognition of the context and perceived emotion of dog barks. 2018 IEEE International Conference on Acoustics, Speech and Signal Processing (ICASSP). pp. 5134-5138. https://doi.org/10.1109/ICASSP.2018.8461757

Van Heijst, B.A.C., De Kort, M.A.C., Overgaauw, P.A.M., Vinke, C.M., Beekmans, M.H.C., 2015. Feiten & cijfers gezelschapsdierensector 2015, Werkgroep Feiten en Cijfers, adviesgroep externe deskundigen. Available at: https://edepot.wur.nl/361828.

Herron, M.E., Buffington, C.A.T., 2010. Environmental enrichment for indoor cats. Compendium: Continuing Education for Veterinarians 32(12): E4.

Hilmer, S., Algar, D., Neck, D., Schleucher, E., 2010. Remote sensing of physiological data: Impact of long term captivity on body temperature variation of the feral cat (*Felis catus*) in Australia, recorded via Thermochron iButtons. Journal of Thermal Biology 35: 205-210. https://doi.org/10.1016/j.jtherbio.2010.05.002

Jackson, B.L., Lehmkuhl, L.B., Adin, D.B., 2014. Heart rate and arrhythmia frequency of normal cats compared to cats with asymptomatic hypertrophic cardiomyopathy. Journal of Veterinary Cardiology 16: 215-225. https://doi.org/10.1016/j.jvc.2014.10.001

Jonckheer-Sheehy, V.S.M., Vinke, C.M., Ortolani, A., 2012. Validation of a Polar® human heart rate monitor for measuring heart rate and heart rate variability in adult dogs under stationary conditions. Journal of Veterinary Behavior: Clinical Applications and Research 7: 205-212. https://doi.org/10.1016/j.jveb.2011.10.006

Jukan, A., Masip-Bruin, X., Amla, N., 2017. Smart computing and sensing technologies for animal welfare: a systematic review. ACM Computing Surveys 50: 10. https://doi.org/10.1145/3041960

Katayama, M., Kubo, T., Mogi, K., Ikeda, K., Nagasawa, M., Kikusui, T., 2016. Heart rate variability predicts the emotional state in dogs. Behavioural Processes 128: 108-112. https://doi.org/10.1016/j.beproc.2016.04.015

Kim, Y., Sa, J., Chung, Y., Park, D., Lee, S., 2018. Resource-efficient pet dog sound events classification using LSTM-FCN based on time-series data. Sensors 18: 4019. https://doi.org/10.3390/s18114019

King, T., Marston, L.C., Bennett, P.C., 2012. Breeding dogs for beauty and behaviour: Why scientists need to do more to develop valid and reliable behaviour assessments for dogs kept as companions. Applied Animal Behaviour Science 137: 1-12. https://doi.org/10.1016/j.applanim.2011.11.016

Komori, Y., Ohno, K., Fujieda, T., Suzuki, T., Tadokoro, S., 2015. Detection of continuous barking actions from search and rescue dogs' activities data. In: 2015 IEEE/RSJ International Conference on Intelligent Robots and Systems (IROS). pp. 630-635. https://doi.org/10.1109/IROS.2015.7353438

Köster, L.S., Sithole, F., Gilbert, G.E., Artemiou, E., 2019. The potential beneficial effect of classical music on heart rate variability in dogs used in veterinary training. Journal of Veterinary Behavior 30: 103-109. https://doi.org/10.1016/j.jveb.2018.12.011

Kreissl, H., Neiger, R., 2015. Measurement of body temperature in 300 dogs with a novel noncontact infrared thermometer on the cornea in comparison to a standard rectal digital thermometer. Journal of Veterinary Emergency and Critical Care 25: 372-378. https://doi.org/10.1111/vec.12302

Krvavica, A., Likar, Š., Brložnik, M., Domanjko-Petrič, A., Avbelj, V., 2016. Comparison of wireless electrocardiographic monitoring and standard ECG in dogs. In: 2016 39th International Convention on Information and Communication Technology, Electronics and Microelectronics (MIPRO). pp. 396-399. https://doi.org/10.1109/MIPRO.2016.7522175

Kuhne, F., Hößler, J.C., Struwe, R., 2014. Behavioral and cardiac responses by dogs to physical human-dog contact. Journal of Veterinary Behavior – Clinical Applications and Research 9: 93-97. https://doi.org/10.1016/j.jveb.2014.02.006

Kumpulainen, P., Valldeoriola, A., Somppi, S., Törnqvist, H., Väätäjä, H., Majaranta, P., Surakka, V., Vainio, O., Kujala, M.V., Gizatdinova, Y., Vehkaoja, A., 2018. Dog activity classification with movement sensor placed on the collar. In: Proceedings of the Fifth International Conference on Animal-Computer Interaction, ACI 2018. Atlanta, Georgia, USA, p. 6. https://doi.org/10.1145/3295598.3295602

Ladha, C., n.d. The future of animal movement science? The Veterinary Health Innovation Engine (vHive). Available at: https://vhive.buzz/future-animal-movement-science/.

Laing, B.Y., Mangione, C.M., Tseng, C.-H., Leng, M., Vaisberg, E., Mahida, M., Bholat, M., Glazier, E., Morisky, D.E., Bell, D.S., 2014. Effectiveness of a smartphone application for weight loss compared with usual care in overweight primary care patients. Annals of Internal Medicine 161: S5-S12. https://doi.org/10.7326/M13-3005

Lemasson, G., Pesty, S., Duhaut, D., Lemasson, G., Pesty, S., Duhaut, D., Lemasson, G., Pesty, S., Duhaut, D., 2013. Increasing communication between a man and a dog. In: Proceedings of the 2013 IEEE 4th International Conference on Cognitive Infocommunications (CogInfoCom'13). HAS Archives, Budapest, Hungary, pp. 145-148.

Mancini, C., 2011. Animal-computer interaction: a manifesto. Interactions 18: 69-73. https://doi.org/10.1145/1978822.1978836

Martin, K.W., Olsen, A.M., Duncan, C.G., Duerr, F.M., 2017. The method of attachment influences accelerometer-based activity data in dogs. BMC Veterinary Research 13: 48. https://doi.org/10.1186/s12917-017-0971-1

Minors, S.L., O'Grady, M.R., 1997. Heart rate variability in the dog: is it too variable? Canadian Journal of Veterinary Research 61: 134-144.

Morrison, R., Reilly, J.J., Penpraze, V., Pendlebury, E., Yam, P.S., 2014. A 6-month observational study of changes in objectively measured physical activity during weight loss in dogs. Journal of Small Animal Practice 55: 566-570. https://doi.org/10.1111/jsap.12273

Naik, R., Witzel, A., Albright, J.D., Siegfried, K., Gruen, M.E., Thomson, A., Price, J., Lascelles, B.D.X., 2018. Pilot study evaluating the effect of feeding method on overall activity of neutered indoor pet cats. Journal of Veterinary Behavior 25: 9-13. https://doi.org/10.1016/j.jveb.2018.02.001

Nederlandse Voedingsindustrie Gezelschapsdieren, 2020. Feiten en cijfers over diervoeding. Feiten en cijfers. Available at: https://www.nvg-diervoeding.nl/over-huisdieren/feiten-cijfers/.

Neidhart, L., Boyd, R., 2002. Companion animal adoption study. Journal of Applied Animal Welfare Science 5: 175-192. https://doi.org/10.1207/S15327604JAWS0503_02

Nuttall, T., McEwan, N., 2006. Objective measurement of pruritus in dogs: a preliminary study using activity monitors. Veterinary Dermatology 17: 348-351. https://doi.org/https://doi.org/10.1111/j.1365-3164.2006.00537.x

Ogata, N., Kikusui, T., Takeuchi, Y., Mori, Y., 2006. Objective measurement of fear-associated learning in dogs. Journal of Veterinary Behavior: Clinical Applications and Research 1: 55-61. https://doi.org/10.1016/j.jveb.2006.06.002

Okin, G.S., 2017. Environmental impacts of food consumption by dogs and cats. PLoS ONE 12: e0181301. https://doi.org/10.1371/journal.pone.0181301

Ortmeyer, H.K., Robey, L., McDonald, T., 2018. Combining actigraph link and petpace collar data to measure activity, proximity, and physiological responses in freely moving dogs in a natural environment. Animals 8: 230. https://doi.org/10.3390/ani8120230

Overbay, T., 2018. The future of the Internet of animals. Global AgInvesting. Available at: https://tinyurl.com/ymcyh6me.

Palestrini, C., Calcaterra, V., Cannas, S., Talamonti, Z., Papotti, F., Buttram, D., Pelizzo, G., 2017. Stress level evaluation in a dog during animal-assisted therapy in pediatric surgery. Journal of Veterinary Behavior 17: 44-49. https://doi.org/10.1016/j.jveb.2016.09.003

Palestrini, C., Previde, E.P., Spiezio, C., Verga, M., 2005. Heart rate and behavioural responses of dogs in the Ainsworth's strange situation: a pilot study. Applied Animal Behaviour Science 94: 75-88. https://doi.org/10.1016/j.applanim.2005.02.005

Parker, M., Lamoureux, S., Allouche, B., Brossier, J.A., Weber, M., Feugier, A., Moniot, D., Deputte, B., Biourge, V., Serra, J., 2017. Accuracy assessment of spatial organization and activity of indoor cats using a system based on ultrawide band technology. Journal of Veterinary Behavior: Clinical Applications and Research 21: 13-19. https://doi.org/10.1016/j.jveb.2017.06.003

Parker, M., Lamoureux, S., Challet, E., Deputte, B., Biourge, V., Serra, J., 2019. Daily rhythms in food intake and locomotor activity in a colony of domestic cats. Animal Biotelemetry 7: 25. https://doi.org/10.1186/s40317-019-0188-0

Patel, S.I., Miller, B.W., Kosiorek, H.E., Parish, J.M., Lyng, P.J., Krahn, L.E., 2017. The effect of dogs on human sleep in the home sleep environment. Mayo Clinic Proceedings 92: 1368-1372. https://doi.org/https://doi.org/10.1016/j.mayocp.2017.06.014

Pavelski, M., Silva, D.M., Leite, N.C., Junior, D.A., De Sousa, R.S., Gu Erios, S.D., Dornbusch, P.T., 2015. Infrared thermography in dogs with mammary tumors and healthy dogs. Journal of Veterinary Internal Medicine 29: 1578-1583. https://doi.org/10.1111/jvim.13597

Petrie, J.P., 2005. Practical application of holter monitoring in dogs and cats. Clinical Techniques in Small Animal Practice 20: 173-181. https://doi.org/10.1053/j.ctsap.2005.05.006

Pongrácz, P., Molnár, C., Miklósi, Á., 2006. Acoustic parameters of dog barks carry emotional information for humans. Applied Animal Behaviour Science 100: 228-240. https://doi.org/ https://doi.org/10.1016/j.applanim.2005.12.004

Pons, P., Jaen, J., Catala, A., 2014. Animal ludens: building intelligent playful environments for animals. In: ACM International Conference Proceeding Series 11-14-Nove, article 3. https://doi.org/10.1145/2693787.2693794

Protopopova, A., Kisten, D., Wynne, C., 2016. Evaluating a humane alternative to the bark collar: automated differential reinforcement of not barking in a home-alone setting. Journal of Applied Behavior Analysis 49: 735-744. https://doi.org/10.1002/jaba.334

Raad voor Dierenaangelegenheden, 2020. Fokken met dieren, voor wie doen we dat? Raad voor Dierenaangelegenheden, Den Haag, the Netherlands, 12 pp. Available at: https://tinyurl.com/xebvt8w8.

Raad voor Dierenaangelegenheden, 2019. Dierbare hulpverleners. Raad voor Dierenaangelegenheden, Den Haag, the Netherlands, 35 pp. Available at: https://tinyurl.com/354sp3ws.

Rawassizadeh, R., Wac, K., Tomitsch, M., 2013. Theme issue on electronic memories and life logging. Personal and Ubiquitous Computing 17: 603-604. https://doi.org/10.1007/s00779-012-0509-2

Rehn, T., Keeling, L.J., 2011. The effect of time left alone at home on dog welfare. Applied Animal Behaviour Science 129: 129-135. https://doi.org/https://doi.org/10.1016/j.applanim.2010.11.015

Semper, H.M., Povey, R., Clark-Carter, D., 2016. A systematic review of the effectiveness of smartphone applications that encourage dietary self-regulatory strategies for weight loss in overweight and obese adults. Obesity Reviews 17: 895-906. https://doi.org/10.1111/obr.12428

Sinski, J., 2016. 'A cat-sized hole in my heart': public perceptions of companion animal adoption in the USA BT. In: Pręgowski, M.P. (ed.) Companion animals in everyday life: situating human-animal engagement within cultures. Palgrave Macmillan, New York, NY, USA, pp. 73-89. https://doi.org/10.1057/978-1-137-59572-0_6

Spangenberg, E.M.F., Björklund, L., Dahlborn, K., 2006. Outdoor housing of laboratory dogs: effects on activity, behaviour and physiology. Applied Animal Behaviour Science 98: 260-276. https://doi.org/https://doi.org/10.1016/j.applanim.2005.09.004

Stampler, L., 2014. There's now facial recognition software for cats. Time, July 16. Available at: https://tinyurl.com/ydf2zkx4.

Steiss, J.E., Schaffer, C., Ahmad, H.A., Voith, V.L., 2007. Evaluation of plasma cortisol levels and behavior in dogs wearing bark control collars. Applied Animal Behaviour Science 106: 96-106. https://doi.org/10.1016/j.applanim.2006.06.018

Strickler, B.L., Shull, E.A., 2014. An owner survey of toys, activities, and behavior problems in indoor cats. Journal of Veterinary Behavior 9: 207-214. https://doi.org/https://doi.org/10.1016/j.jveb.2014.06.005

The European Pet Food Industry, 2019. Facts & figures 2019 European overview. The European petfood industry, Brussels, Belgium.

Thomas, D.G., Post, M., Bosch, G., 2017. The effect of changing the moisture levels of dry extruded and wet canned diets on physical activity in cats. Journal of Nutritional Science 6: E9. https://doi.org/10.1017/jns.2017.9

Thomas, R.L., Baker, P.J., Fellowes, M.D.E., 2014. Ranging characteristics of the domestic cat (*Felis catus*) in an urban environment. Urban Ecosystems 17: 911-921. https://doi.org/10.1007/s11252-014-0360-5

Thompson, R.J., 2018. The use of wearable sensors for animal behaviour assessment. PhD thesis Newcastle, UK, 159 pp. Available at:https://core.ac.uk/download/pdf/334997949.pdf.

Travain, T., Colombo, E.S., Grandi, L.C., Heinzl, E., Pelosi, A., Prato Previde, E., Valsecchi, P., 2016. How good is this food? A study on dogs' emotional responses to a potentially pleasant event using infrared thermography. Physiology and Behavior 159: 80-87. https://doi.org/10.1016/j.physbeh.2016.03.019

Trendpanel Gezelschapsdieren, 2019. Pet monitor 2019. HAS Hogeschool, Aeres Hogeschool, Den Bosch, Dronten, the Netherlands.

Vailshery, L.S., 2021. Connected wearable devices worldwide 2016-2022. Statista Inc., New York, NY, USA. Available at: https://tinyurl.com/2s64yv8z

von Borell, E., Langbein, J., Després, G., Hansen, S., Leterrier, C., Marchant-Forde, J., Marchant-Forde, R., Minero, M., Mohr, E., Prunier, A., Valance, D., Veissier, I., 2007. Heart rate variability as a measure of autonomic regulation of cardiac activity for assessing stress and welfare in farm animals – a review. Physiology and Behavior 92: 293-316. https://doi.org/10.1016/j.physbeh.2007.01.007

Waddell, R.E., Marino, D.J., Loughin, C.A., Tumulty, J.W., Dewey, C.W., Sackman, J., 2015. Medical infrared thermal imaging of cats with hyperthyroidism. American Journal of Veterinary Research 76: 53-59. https://doi.org/10.2460/ajvr.76.1.53

Wang, P., Ma, Y., Liang, F., Zhang, Y., Yu, X., Li, Z., An, Q., 2020. Non-contact vital signs monitoring of dog and cat using a UWB radar. Animals 10: 205. https://doi.org/10.3390/ani10020205

Wormald, D., Lawrence, A.J., Carter, G., Fisher, A.D., 2017. Reduced heart rate variability in pet dogs affected by anxiety-related behaviour problems. Physiology and Behavior 168: 122-127. https://doi.org/10.1016/j.physbeh.2016.11.003

Yamazaki, A., Edamura, K., Tanegashima, K., Tomo, Y., Yamamoto, M., Hirao, H., Seki, M., Asano, K., 2020. Utility of a novel activity monitor assessing physical activities and sleep quality in cats. PLoS ONE 15: e0236795. https://doi.org/10.1371/journal.pone.0236795

Zanghi, B.M., 2016. Eye and ear temperature using infrared thermography are related to rectal temperature in dogs at rest or with exercise. Frontiers in Veterinary Science 3: 111. https://doi.org/10.3389/fvets.2016.00111

Zupan, M., Buskas, J., Altimiras, J., Keeling, L.J., 2016. Assessing positive emotional states in dogs using heart rate and heart rate variability. Physiology and Behavior 155: 102-111. https://doi.org/10.1016/j.physbeh.2015.11.027

8. Precision technology and sensors

E. van Erp-van der Kooij

HAS University of Applied Sciences, P.O. Box 90108, 5200 MA 's Hertogenbosch, the Netherlands; l.verp@has.nl

Highlights

- Sensors measure and quantify observations.
- The output signal of a sensor is a signal that can be stored.
- Sensors can be handheld, attached to an animal or mounted in the animal pen.
- Examples of sensors are microphones, cameras and accelerometers.

8.1 Precision technology

Precision technology is derived from precision engineering, which is a subdiscipline of several other engineering areas such as software, electronics and mechanics. Precision engineering is defined as working with machines and devices with exceptionally low variation in the measured value, high repeatability and high stability over time. This results in advantages such as creating very precise movement (e.g. in robots); making (automatic) assembly easier; reducing the initial and running costs; extending the life span; improving interchangeability of components (so that corresponding parts made by other factories or firms can be used in their place); improving quality control through higher machine accuracy capabilities (reducing conventional inspection); achieving a longer life of components (due to less wear/fatigue); making functions independent of one another; achieving greater miniaturisation and packing densities and achieving further advances in technology (Venkatesh and Izman, 2008). In many industries precision technology has been introduced, also in agriculture and companion animals. In farming, precision technology is called precision agriculture, precision farming or 'smart farming'. Precision Livestock Farming is derived from that term. Precision technology has found a place in the consumer market for human health, with many wearable sensors that people use in sports and leisure. In line with that interest, the market for sensors and devices for companion animals has also grown considerably, and many devices and technologies have been introduced for dogs, cats and horses.

8.2 Sensors

Sensors are used to measure and quantify observations. Sensors convert measured phenomena into a quantity, which can be stored. The output of a sensor is a signal that can be biological, chemical, electric or electromagnetic and many more. There are a lot of different sensors available. Some sensors are attached to objects or animals, such as activity sensors based on mechanical motion (e.g. acceleration). Some remote sensors use optical information, such as cameras used for estimating grass production, to record animal behaviour or to identify animals. Thermal sensors can detect inflammation or stress by measuring body temperature. Sensors can be used to measure milk quality and those can be based on optical, mechanical, thermal or electrical information. This large variety of sensors also have a lot in common. Sensor measurements usually result in a continuous flow of electrical signals, that have to be interpreted and translated into output. Depending on the measurement frequency, several measurements per second, per hour or per day can be recorded (Lokhorst, 2018). Data can be used to visualise measurements or to analyse and apply algorithms, to translate data into information. This data analysis or data science is described in Chapter 9. In this chapter, the working mechanism and relevant applications of several sensors are described.

8.2.1 Acoustic sensors

Sound is usually captured with a microphone, which is a device that converts sounds into an electric signal. Sound waves are air pressure variations, that can be visualised in a spectrogram (Bishop *et al.*, 2019). There are four main types of microphones: condenser microphones, dynamic or moving-coil microphones, ribbon microphones and piezo microphones. MEMS microphones ('mic on a chip') are based on the same principles. Microphones can be used to monitor animal sound patterns in groups of animals or specific sounds in individual animals, related to production and health (Halachmi *et al.*, 2019; Neethirajan *et al.*, 2017). Sound can also be used to detect malfunctioning machines or devices on the farm (Lee *et al.*, 2019; Lüttenberg *et al.*, 2018).

The *condenser microphone* was invented in 1916 by Wente. A condenser microphone, or capacitator, uses two thin metal plates, very close to each other. One of the plates is movable and behaves as a diaphragm, while the other plate, the backplate, is fixed. The moving plate is generally gold plated to conduct electrical current properly. The plates are connected to a converter that generates output tension. As the plate moves it creates an electric potential difference. This captured signal is amplified and transmitted. The internal circuit of the microphone increases the gain of the signal and requires electrical power to work. The metal plates also need to be powered with electrical tension. Condenser microphones can be unidirectional, bidirectional or omnidirectional (Huffman, n.d.), depending on whether it captures sound from one direction, two directions or all directions. The *dynamic microphone* uses a small movable induction coil, positioned in the magnetic field of a permanent magnet, which is attached to the diaphragm. When sound enters through the windscreen of the microphone, the sound wave moves the diaphragm. When the diaphragm vibrates, the coil moves in the magnetic field, producing a varying current in the coil through electromagnetic induction. Dynamic microphones are robust, relatively inexpensive and resistant to moisture (Marshall and Harry, 1939; Yoshida, 1984). The next type of microphone is the *ribbon microphone*, invented in 1924 by Scottky. Ribbon microphones use a ribbed aluminium membrane at right angles to a strong magnetic field. The audio frequency current that passes through this membrane causes it to move and create sound vibrations. Ribbon microphones are fragile because of the thin aluminium ribbon, but the newest models use nanomaterials which overcome this problem. Ribbon microphones are the oldest design still used at major music studios. Ribbon microphones are unidirectional or bidirectional (Dooley, 2014; Viscarolasaga, 2008). *Crystal microphones* or *piezo microphones* use the phenomenon of piezoelectricity, which is the ability of some materials to produce a voltage when subjected to pressure. Piezo microphones convert vibrations into electrical signals. Piezoelectric transducers are often used as contact microphones to amplify sound from acoustic musical instruments and to record sound in challenging environments, such as underwater under high pressure (Lee and Lee, 2008). The *MEMS* or Micro Electrical Mechanical System microphone is also called a microphone chip or silicon microphone. A pressure-sensitive diaphragm is etched directly into a silicon wafer by MEMS processing techniques and is usually accompanied by an integrated preamplifier. Most MEMS microphones are variants of the condenser microphone design, but there are also piezoelectric MEMS microphones. Digital MEMS microphones have built-in analog-to-digital converter (ADC) circuits on the same chip, making the chip a digital microphone. This way it can be easily integrated with modern digital products (Johnson, 2014).

8.2.2 Chemical sensors

A chemical sensor can measure (part of) the chemical composition of the liquid or gas environment. The sensor first has to recognise a certain chemical compound, and second, give information on the concentration of this compound. To recognise a compound, molecules interact selectively with receptor molecules or with specific sites in the sensor. Next, a physical parameter (e.g. colour) will vary and this variation is translated into an output signal. If the recognition of the compound by the sensor is based on a biological principle, the sensor is called a biosensor (Banica, 2012). In livestock, chemical sensors are used in milk analysis or breath analysis in dairy cows (Küntzel et al., 2018), for example, to measure progesterone in dairy milk for heat or pregnancy detection. Chemical sensors can also be used to monitor climate conditions in animal housing by measuring ammonia or carbon dioxide (Fournel et al., 2017; Hayes et al., 2006).

8.2.3 Electrical current and conductivity

Current sensing is the technique to measure electric current, ranging from picoamperes to thousands of amperes. The selection of the method of current sensing depends on factors such as magnitude, accuracy, bandwidth, robustness, isolation, size and costs. Some devices directly display the current, or the current can be converted to a digital signal to use to monitor or control a system (Costa et al., 2001). In farms, current sensing is mostly used in machine maintenance (Lüttenberg et al., 2018). In animal science, milk conductivity is an interesting parameter for dairy cow health. Conductivity can be measured by placing two electrodes in a fluid with a known volume, and the conductivity of that fluid is compared to the conductivity of a standard NaCl solution. Milk from cows with mastitis show altered concentrations of Na^+, K^+ and Cl^-, resulting in increased conductivity. Therefore conductivity of milk, measured with a measuring cell placed in the milking equipment, is used as an indicator of mastitis (Nielen et al., 1992).

8.2.4 Flow

A mass flow meter measures the flow rate of a fluid through a tube. The output is mass per time unit. To calculate the volume per unit of time, the mass flow rate is divided by the fluid density (Henry *et al.*, 2000). The volumetric flow rate can also be determined based on the conductivity of the fluid. This is done by placing two electrically conducting elements in-line, which communicate with an electronic circuit. The quantity of the liquid that has flowed by is determined by the time interval, the distance between the conducting elements, the flow rate and the electrical resistance. Flow meters are used in milking machines for dairy cows to determine the flow rate, the milk yield per quarter and per cow; the decrease in flow rate determines the timing of the automatic removal of the teat cups from the cow's teats (Ordolff, 2001; Van den Berg, 1998).

8.2.5 Mechanical sensors

Mechanical sensors can measure position, angle or acceleration. Broadly used in livestock and companion animals are accelerometers. An *accelerometer* measures acceleration: the change in velocity in time (Watson, 1986). The concept of an accelerometer is that of a mass on a spring, the 'proof mass'. When the accelerometer accelerates, the mass is moved and the spring is compressed. This compression is a measure of acceleration. Mechanical accelerometers usually have an electronic circuit that senses motion and pushes on the proof mass while an electromagnet keeps the mass from moving far. Modern accelerometers are often small MEMS (micro-electro-mechanical systems), consisting of a cantilever beam with a proof mass. When there is acceleration, the deflection of the proof mass from its neutral position is measured by using the capacitance between a set of beams attached to the proof mass and a set of fixed beams. Capacitance is measured by the amount of electric charge stored on a conductor per unit difference in electrical potential. Capacitance is measured in F (farad), which is coulomb (unit of electric charge) per volt (unit of potential difference) (Seiko Suzuki *et al.*, 1992). Single- and multi-axis accelerometers can detect the magnitude and direction of acceleration and can be used to measure orientation, vibration and shock. The position of the sensor can be determined accurately when the sensor is static; for moving sensors, the orientation of the device is needed to calculate the position. The precise movement can be calculated if sufficient sample rates are used.

Accelerometers attached to animals have become increasingly popular since battery life has increased, and sensors are now small enough to be attached to animals' legs, necks, ears, or tails. It is even possible to attach an accelerometer to the dorsal fin of fish. In practice, high sampling rates are required for accelerometers to acquire useful data, and this sending of data with a high frequency limits battery life. One solution for this problem would be, to not send the raw data, but to make calculations within the sensor. This means that only simple calculations can be done, and no complex algorithms to translate data into information can be used. With evolving battery technology and decreasing power consumption of electronic devices, this problem will become smaller and more complex calculations can be done within the sensor. In livestock and companion animals, accelerometers are used to record behaviours and behavioural patterns from activity data (Halachmi et al., 2019; Lavanaya and Mallappa, 2019). Accelerometers are also used in machine maintenance. By measuring vibration in for example fans or feeding machines, anomalies can be detected and preventive or predictive maintenance can be performed, before machines break down (Campagnie, 2021; Lee et al., 2019; Lüttenberg et al., 2018).

A *gyroscope* measures angle. Many activity sensors contain an accelerometer as well as a gyroscope (Halachmi, 2015; Luinge et al., 1999; Watson, 1986). A gyroscope measures the angle of the sensor, which can be used to detect body posture or limb posture, e.g. to detect lameness in horses (Keegan et al., 2002), lying versus standing, and lameness in cows (O'Leary et al., 2020) or to predict calving based on the tail position of the cow (Krieger et al., 2018). A gyroscope is used for measuring or maintaining orientation, based on the principles of angular momentum. A conventional gyroscope (Figure 8.1) is a mechanism comprising of a rotor made to spin about one axis, the screws of the rotor being mounted in an inner ring or 'gimbal', and the inner ring being fixed for oscillation into an outer ring which is in turn fixed for oscillation to a relative support. The outer ring is mounted, so it can turn about an axis in its plane, determined by the support. The gyroscope is a spinning wheel or disk whose axle is free to take any orientation or direction. Because the gyroscope spins very fast, this creates momentum, and therefore the direction changes much less in response to a known external force or 'torque' than it would when it was not spinning.

An *inclinometer*, also called a tilt sensor, slope sensor, or clinometer is a sensor that measures the angle of an object, using the force of gravity. Inclinometers determine the pitch or roll angle and translate this into an electrical output signal (FRABA B.V., n.d.). Inclinometers can be used to measure tail-position in cows to predict calving (Voss et al., 2020).

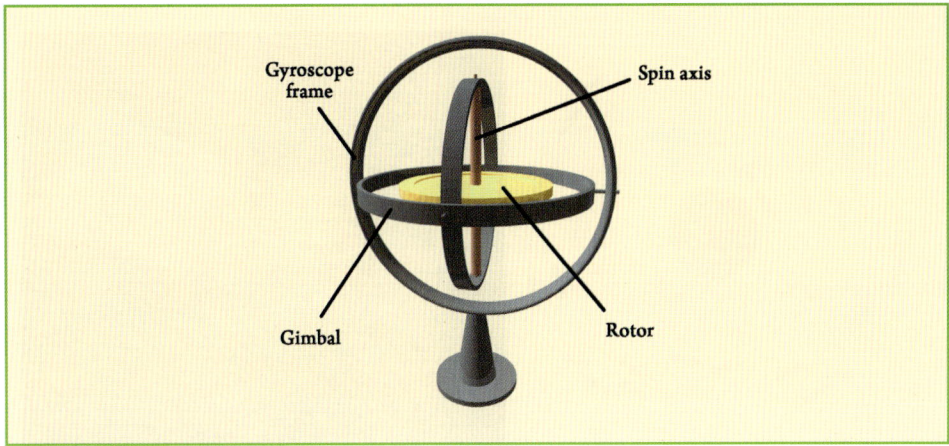

Figure 8.1. Gyroscope (wikicommons, L. Vieira).

8.2.6 Location

Location can be measured in several ways. Outdoors, a *global positioning system* (GPS) system can be used (Kaplan and Hegarty, 2006). GPS was designed by the USA military in the late '60s. The first satellite was launched in 1978 and the first hand-held GPS receiver was introduced in 1989. Since 1996, GPS is free to use by civilians (Maddison and Ni Mhurchu, 2009). A tracking system using GPS satellites provides extremely accurate position, velocity and time information for vehicles or other objects within any mobile information system (Brown and Sturza, 1993). The GPS is a US-owned utility, consisting of the space segment, with 24 operating satellites that transmit one-way signals that give the current GPS satellite position and time (Figure 8.2); the control system, consisting of monitor and control systems that maintain the satellite in their proper orbits, tracking the GPS satellites and uploading the updated navigational data; and the user segment, consisting of the GPS receiver equipment receiving the signals from the GPS satellites and using the transmitted information to calculate the position and time (United States Government, n.d.). A tracking system includes a sensor attached to the object, animal or person, a communication link, a workstation and a GPS reference receiver. Three satellite measurements are required to calculate the location, also referred to as triangulation. GPS tracking systems can not only be used to track animals but also to determine their behaviour, based on position; in outdoor cows, grazing, resting and walking could be determined from GPS data (Schlecht *et al.*, 2004).

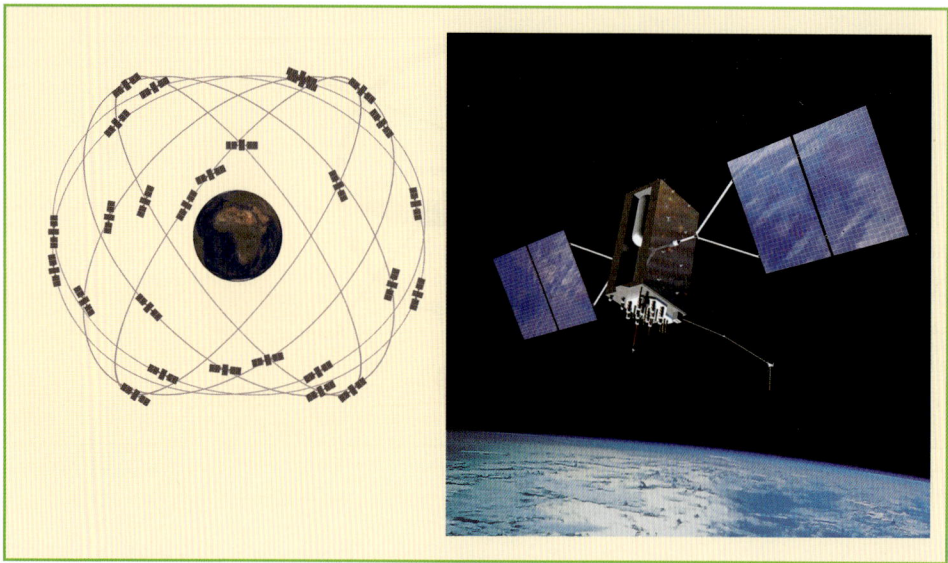

Figure 8.2. (A) Constellation of 24 GPS satellites, each circling Earth twice a day (source: United States Government, www.gps.gov/multimedia/images); and (B) GPS III satellite (source: United States Government, www.gps.gov/multimedia/images).

Indoors, especially inside animal farms with metal fences and feeding racks, the use of GPS can be inaccurate (Kjærgaard *et al.*, 2010). However, there are several other options to track animals indoors.

For indoor tracking, *radio frequency identification* (RFID) tags can be used. When an animal wearing an RFID tag is close enough to a reader, the location is measured. A passive RFID tag does not need a power source, because it uses an antenna to produce the needed energy (Huhtala *et al.*, 2007).

Radio tracking is another possibility. In this system, antennae are positioned at fixed points in the field or barn, which forward signals from tagged animals to the system controller that collects the data. Through triangulation, the location of each animal is calculated. This method is used in the cow positioning system of Nedap (Huhtala *et al.*, 2007).

Bluetooth and *GSM* devices can be used for positioning by giving a signal when another device is close enough. Based on which locator beacons the app can detect, combined with the strength of the signal from each of the beacons and their known location, triangulation is used to calculate the current location. The range for Bluetooth signals is approximately 10 meter, so devices should be in that range (Huhtala et al., 2007).

A wireless local area network such as *Wi-Fi* can be used to determine location by calculating the time difference of arrival (TDOA). The location receiver receives standard messages and uses an algorithm based on TDOA to calculate the location of the tags (Huhtala et al., 2007). Based on the same principle, location data can be calculated from Ultra Wide Band (*UWB*) radio signals. With tags transmitting extremely short UWB pulses and remote sensors, location can be mapped by using TDOA and angle of arrival (AoA). This method is applied in the Tracklab system using Ubisense UWB sensors; this can be applied to track dairy cows in a barn (Frondelius et al., 2014; Pastell et al., 2018).

8.2.7 Optical, imaging, light

To determine levels of light, light intensity loggers or radiation sensors can be used (Long et al., 2012). A light sensor or electro-optical sensor is a photoelectric device that converts light energy or photons to electrical energy or electrons. Light is measured in Lumen, which is the total amount of visible light from a light source, or in Lux, which is the total amount of light that falls on a particular surface. There are three types of light sensors: photoresistors, photodiodes and phototransistors. *Photoresistors* or light-dependent resistors (LDR) are the most common light sensors. Photoresistors consist of cadmium sulphide cells, that are sensitive to visible and near-infrared light. High light intensity causes a lower resistance between the cadmium sulphide cells, while low light intensity causes a higher resistance between the cells. Photoresistors are used in streetlamps that automatically turn on when it gets dark. *Photodiodes* consist of silicon and germanium materials and have optical filters and built-in lenses. When light hits a photodiode, electrons are loosened, resulting in an electrical current. With brighter light, the electrical current is stronger. Photodiodes are used in CD players, smoke detectors, medical measuring equipment and solar panels. In livestock, photodiodes are used in heartbeat sensors. The heartbeat sensor measures the change in volume of blood in any part of the animal body which results in a variation in the light intensity. The flow of blood is set by the speed of heartbeat and since light is absorbed by blood, the signal pulses are equal to that of the beating heart. A simple heartbeat sensor consists of an LED and a photodiode that detects light (Nanda et al., 2020). Another application of photodiodes is in online equipment to measure progesterone in the milk of dairy cows. The photodiode is part of the biosensor that can recognise specific molecules such as hormones. Using an enzyme immune-assay where more progesterone in the sample results in fewer light units, photodiodes are used to measure these light units, which is used to calculate the progesterone content of the milk (Mottram, 2016).

Phototransistors consist of a photodiode with an amplifier, which increases the light sensitivity of the phototransistor. The working principle is similar to that of a photodiode (Shawn, 2020). Light sensors can be used to measure light in closed farm systems, such as lighting schedules in broiler farms or level of light in pig farms, but also to recognise plants or animals. In these applications, phototransistors are used to recognise different colours, and with that information, algorithms are used to detect fruit in fruit trees or to identify animals (Rosell-Polo *et al.*, 2015). An example of an application for a photo-transistor circuit is a system for the automatic removal of teat cups in a milking device for a dairy cow. The phototransistor detects the milk-air interface in the milk flow indicator; when the milk flow rate is below a certain threshold, the teat cups are automatically removed (Nichols, 1972).

8.2.8 Cameras

Optical sensors form the basic technology of cameras, vision and image recognition systems, which are widely used in precision livestock farming and companion animal care (Benjamin and Yik, 2019; Rowe *et al.*, 2019). Examples in livestock are body condition score or weight estimation of pigs and dairy cows, facial recognition to identify individual animals, and detection of behaviours to monitor health and welfare. For dogs and cats, facial recognition can be used to find lost animals and to identify pets for automatic feeding systems.

Cameras detect visible light and form an image. The range of detection of a common camera is 400-700 nanometre. A camera collects and focuses the light. Historically, the film was made up of light-sensitive materials. When those materials were hit with light from the lens, they captured the object and the amount of light that was reflected. In the darkroom, the film that had been exposed to light was put in chemical baths to create the image. Digital cameras work differently. The sensor in a digital camera resembles a solar panel. The light sensor inside is usually a charge-coupled device or CCD. This is a grid of many little squares that work as solar cells, converting light energy into electrical energy. The light conditions generate specific electrical charges, that can be interpreted by the software in the camera. Each sensor is divided into red, green and blue pixels ('megapixels'). When light hits the pixel, it is converted by the sensor into energy. A computer inside the camera reads the amount of energy produced. This allows the sensor to determine the dark and light image areas along with the colour since each pixel has a colour value. Large sensors perform better in darker conditions and cameras with more megapixels generally show more detail (CreativeLive, n.d.; McDowell, 2009).

8.2.9 Pressure

A pressure sensor consists of a thin, flexible membrane that can deform and that covers a reference cavity that is sealed at low vacuum pressure. When the pressure outside of the membrane is higher than inside the reference cavity, the membrane will stretch or deform into the cavity (Figure 8.3). The deformation of the membrane causes two mechanical changes that we can use for transduction to an electrical measurement: (1) by measuring the change in capacity, which occurs because the membrane moves to the bottom of the cavity; or (2) by measuring the strain on the membrane material with a resistive circuit. To measure the displacement using capacitance, electrodes are built into the membrane and the bottom of the reference cavity. The electrodes act as a parallel plate capacitor. When the membrane deforms, the distance between the membrane and the bottom of the cavity decreases, resulting in increased capacitance. The change in capacitance can be translated into a change in voltage, which can be measured and turned into a digital signal. The second method is to measure the strain in the membrane. When the membrane is stretched, the strain will cause a changed resistance of the material. Semiconductor materials show a large change in resistance due to strain, this phenomenon is called piezoresistance. Many modern pressure sensors use piezoresistive materials to convert the mechanical force into electrical changes that can be turned into a digital signal. Pressure sensors

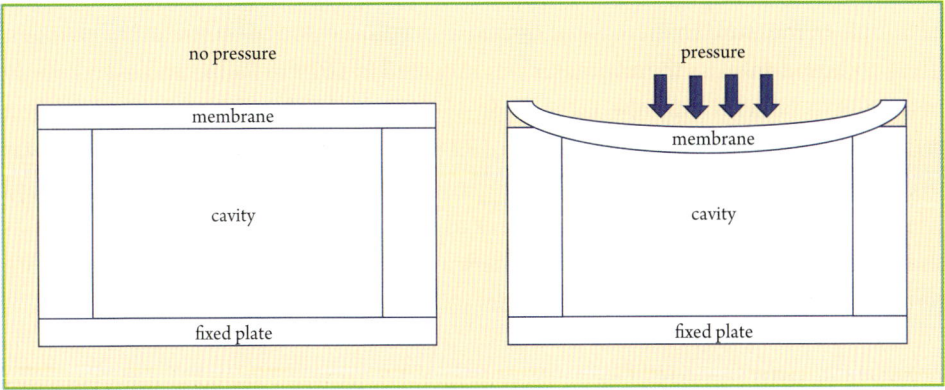

Figure 8.3. Pressure sensor.

were among the first sensors to be miniaturised and mass-produced cheaply in microelectromechanical systems (MEMS) (Clifton, 2019; Fraden, 2016).

Pressure sensors are used in weighing devices for livestock and companion animals (Muddle, 1984) and can be used in some specific applications, for example, to determine rumination and eating in dairy cows with a noseband (Zehner *et al.*, 2017) or to determine lameness in dairy cows using a pressure-sensitive walkway (Maertens *et al.*, 2011).

8.2.10 Temperature sensors

Temperature sensors are used to measure climate conditions outdoors or in the animals' housing system, or the body temperature of individual animals. Body temperature can be measured on the skin of the animal using a thermographic camera, on the outside of the ear with an ear tag, or internally using a rectal or vaginal thermometer. Temperature sensors can be implanted, and for dairy cows, a temperature sensor can be used in a rumen bolus. Temperature can be measured either mechanical or electrical, or based on infrared radiation.

Mechanical temperature sensors are thermometers or bimetallic strips. The first *thermometers* were invented probably more than 2,000 years ago and were based on the principle of fluids or gases expanding and contracting with higher or lower temperatures. Instruments were made with a closed tube partially filled with air and its end in a water container. With increasing temperatures, the air would expand and cause the air to move along the tube, showing the temperature increase, and the other way around with decreasing temperatures. Because these thermometers were sensitive to air pressure, they were not very reliable. The next step was the use of liquids to measure an increase or decrease of volume due to temperature changes. The first liquid-in-glass thermometer was invented in 1629 in Padua and used brandy in a sealed tube with a numbered scale. In 1714 Daniel Fahrenheit invented the first reliable thermometer using mercury. In 1724, the Fahrenheit temperature scale was proposed, and in 1742 Anders Celsius proposed another scale, that is nowadays used in an adjusted form, with 0 degrees as the freezing point and 100 degrees as the boiling point (Middleton, 1966). Thermometers are still widely used to measure temperature in the farm or animal house.

A *bimetallic* strip consists of two different metal parts that expand at different rates when they are heated. Usually, steel and copper are used. If the double strip is heated, it bends one way and if cooled it bends the other way; the metal that expands the most if heated, will be on the outer side when heated and on the inner side when cooled (Figure 8.4) (Sobel, 1995). Bimetallic strips can be used to record air temperature in farms (Stafford and Collison, 1987), but are mostly replaced by other, more modern sensors.

Electrical temperature sensors are thermistors, thermocouples, resistance thermometers or silicon bandgap sensors. The working mechanism is based on the thermoelectric effect of a conductor. If one end of a conductor is placed in a cold area and the other in a warm area, energy will flow from the warm to the cold part. The energy or heat flow is proportional to the thermal conductivity of the conductor. The thermal gradient also sets an electric field inside the conductor, which results in an increasing voltage. A *thermistor* is a thermal resistor. Thermistors are usually metal-oxide sensors and are absolute temperature sensors. The resistance of thermistors changes with an increase in temperature. Since this relation is highly non-linear, a mathematical model is needed to relate the resistance of a thermistor to its absolute temperature. In a thermocouple, two different conductors are joined to form a junction. The voltage produced by the thermocouple depends on the temperature difference between the two thermocouple junctions, making it a relative sensor. To measure temperature with a thermocouple, one junction serves as a reference, with its absolute temperature being measured by a separate absolute sensor. A *thermocouple* generates a voltage in response to temperature and therefore does not require an external power source: it is a passive sensor. Metals or alloys used to make thermocouples are for example Copper versus Constantan or Iron versus Constantan. Constantan is an alloy of Copper and Nickel (Fraden, 2016). *Resistance thermometers* or resistance temperature detectors (RTD) are metal sensors in the form of a wire or thin film. The resistance of these metal

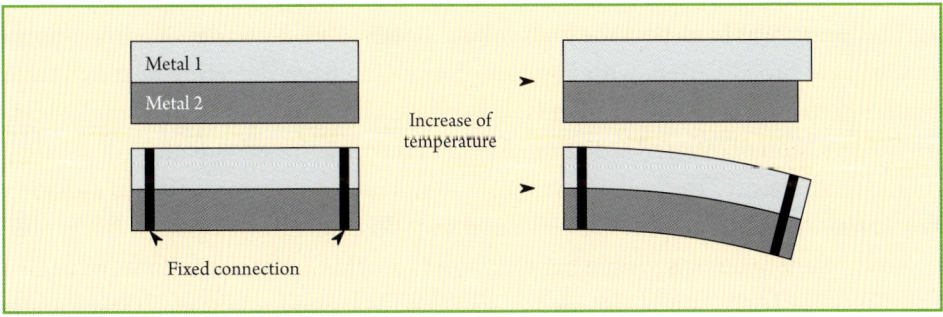

Figure 8.4. Measuring temperature with a bimetallic strip. Source: By Virtual gamma, CC BY-SA 3.0, https://commons.wikimedia.org/w/index.php?curid=12871808.

sensors is dependent on temperature, which makes them suitable for temperature sensing. Platinum is used most often because it is predictable in response, stable and durable (Fraden, 2016). *Silicon bandgap temperature sensors* are silicon diodes that produce a voltage, dependent on temperature. In metals, electrons can easily migrate through the material, therefore electrical conductivity is high. In isolators or semiconductors, such as silicon diodes, electrons must first cross the energy bandgap, therefore conductivity is much lower. In a silicon bandgap sensor, the voltage can be measured with an electronic circuit, and from that voltage, the temperature of the diode is calculated. Silicon bandgap temperature sensors are commonly used in electronic devices and are stable in extreme environmental conditions, due to the stability of crystalline silicon (AZoSensors, 2014; Fraden, 2016).

Thermistors can be implanted to measure body temperature for research purposes, for example in dairy cows (Bitman *et al.*, 1984); however, with the development of non-invasive sensor boluses, this application will be used less often. Internal temperature sensors are usually thermistors, used for example in sensor boluses for dairy cows; for monogastric animals such as horses, dogs or laying hens, thermistors can be ingested and will measure during the time it takes to travel through the digestive tract (Sellier *et al.*, 2014). Thermistors can be used to measure external body temperature, for example in dairy cows (Hoffmann *et al.*, 2020). Silicon bandgap temperature sensors can be used in loggers to record the environmental temperature; they can measure temperature accurately in open air, caves, soil and river environments (Burlet *et al.*, 2015).

Thermographic cameras create images based on *infrared radiation*. Already in 1833, an instrument was invented that could detect a person at 10 meters distance based on detecting infrared radiation. In 1901, a cow could be detected at 400 meters distance. In 1965, the first commercial thermal imaging camera was sold. The first application of this camera was to inspect the high voltage power lines. Nowadays thermographic cameras are broadly used, with applications ranging from detecting heat loss from buildings to animal health problems in animals. The working mechanism depends on the emission of infrared radiation. All objects emit infrared radiation, depending on the temperature. The higher an object's temperature, the more infrared radiation is emitted. A thermographic camera can detect this radiation, similar to the way an ordinary camera detects visible light. Infrared cameras are sensitive to wavelengths from 1000 to 14,000 nm. (Hovinen *et al.*, 2008; Martín Ocaña *et al.*, 2004; Randle *et al.*, 2017). Thermographic cameras are used to measure leg, paw and hoof temperatures to detect lameness in dairy cows, dogs or horses; to monitor temperature of piglet nests; to measure skin temperature as a measure of stress in laying hens; or to measure temperature of chicken and duck eggs, in order to determine their oxygen rate and the incubation rate of the eggs (Herborn *et al.*, 2015; Mortola *et al.*, 2015; Nääs *et al.*, 2014; Stewart *et al.*, 2005). In Figure 8.5A, a thermographic image and in Figure 5B, a normal image of a broiler chicken are shown.

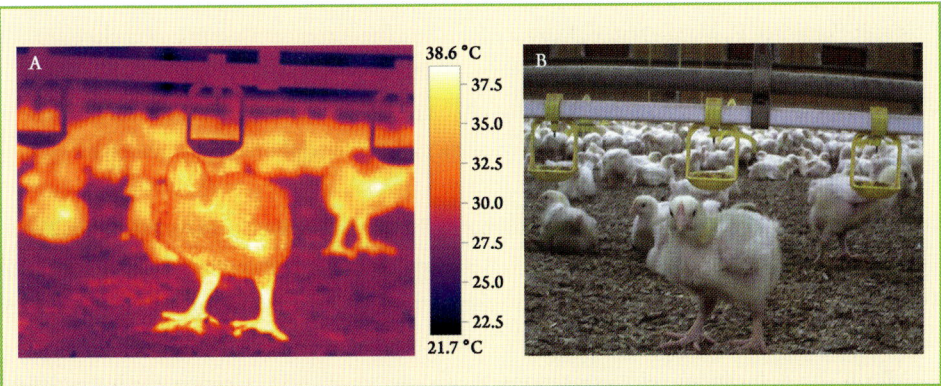

Figure 8.5. Thermographic (A) and normal (B) image of broiler chickens in a Dutch farm (photos: HAS University of Applied Sciences).

References

AZoSensors, 2014. An introduction to silicon bandgap temperature sensors. Available at: https://www.azosensors.com/article.aspx?ArticleID=369.

Banica, F.G., 2012. Chemical sensors and biosensors: fundamentals and applications. John Wiley & Sons, Chicester, UK.

Benjamin, M., Yik, S., 2019. Precision livestock farming in swine welfare: a review for swine practitioners. Animals 9: 133. https://doi.org/10.3390/ani9040133

Bishop, J.C., Falzon, G., Trotter, M., Kwan, P., Meek, P.D., 2019. Livestock vocalisation classification in farm soundscapes. Computers and Electronics in Agriculture 162: 531-542. https://doi.org/10.1016/j.compag.2019.04.020

Bitman, J., Lefcourt, A., Wood, D.L., Stroud, B., 1984. Circadian and ultradian temperature rhythms of lactating dairy cows. Journal of Dairy Science 67: 1014-1023. https://doi.org/10.3168/jds.S0022-0302(84)81400-9

Brown, A.K., Sturza, M.A., 1993. Vehicle tracking system employing global positioning system (GPS) satellites. United States Patent. Available at: https://patents.google.com/patent/US5225842A/en.

Burlet, C., Vanbrabant, Y., Piessens, K., Welkenhuysen, K., Verheyden, S., 2015. Niphargus: A silicon band-gap sensor temperature logger for high-precision environmental monitoring. Computers and Geosciences 74: 50-59. https://doi.org/10.1016/j.cageo.2014.10.009

Campagnie, B., 2021. Choose the right accelerometer for predictive maintenance. Analog Devices, Inc., Norwood, MA, USA. Available at: https://tinyurl.com/5f2nvdjs.

Clifton, 2019. Introduction to pressure sensors. Pocket Lab, Myriad Sensors, Inc., San Jose, CA, USA. Available at: https://tinyurl.com/yzrt86nv.

Costa, F., Poulichet, P., Mazaleyrat, F., Labouré, E., 2001. The current sensors in power electronics, a review. EPE Journal 11: 7-18. https://doi.org/10.1080/09398368.2001.11463473

CreativeLive, n.d. How does a camera work? The ultimate guide to learning photography. CreativeLive, Seattle, WA, USA. Available at: https://www.creativelive.com/photography-guides/how-does-a-camera-work.

Dooley, W.L., 2014. Ribbon microphones. Journal of the Acoustical Society of America 136: 2130.

Fournel, S., Laberge, B., Rousseau, A.N., 2017. Rethinking environment control strategy of confined animal housing systems through precision livestock farming. Biosystems Engineering 155: 96-123. https://doi.org/10.1016/j.biosystemseng.2016.12.005

FRABA B.V., n.d. Inclinometer specifications. Inclinometers. FRABA, Heerlen, the Netherlands. Available at: https://tinyurl.com/72em6tvv.

Fraden, J., 2016. Handbook of modern sensors, 5th ed. Springer International Publishing AG, Basel, Switzerland.

Frondelius, L., Pastell, M., Mononen, J., 2014. Validation of the TrackLab positioning system in a cow barn environment. In: Spink, A., Van den Broek, E.L., Loijens, L., Woloszynowska-Fraser, M., Noldus, L. (eds.) Measuring Behavior 2014. Proceedings of the 9th International conference on methods and techniques in behavioral research. Noldus, Wageningen, the Netherlands, pp. 27-30. Available at: https://tinyurl.com/jds6xzpf.

Halachmi, I. (ed.), 2015. Precision livestock farming applications. Wageningen Academic Publishers, Wageningen, the Netherlands. https://doi.org/10.3920/978-90-8686-815-5

Halachmi, I., Guarino, M., Bewley, J., Pastell, M., 2019. Smart animal agriculture: application of real-time sensors to improve animal well-being and production. Annual Review of Animal Biosciences 7: 403-425. https://doi.org/10.1146/annurev-animal-020518-114851

Hayes, E.T., Curran, T.P., Dodd, V.A., 2006. Odour and ammonia emissions from intensive pig units in Ireland. Bioresource Technology 97: 940-948. https://doi.org/10.1016/j.biortech.2005.04.023

Henry, M.P., Clarke, D.W., Archer, N., Bowles, J., Leahy, M.J., Liu, R.P., Vignos, J., Zhou, F.B., 2000. A self-validating digital Coriolis mass-flow meter: an overview. Control Engineering Practice 8: 487-506. https://doi.org/10.1016/S0967-0661(99)00177-X

Herborn, K.A., Graves, J.L., Jerem, P., Evans, N.P., Nager, R., McCafferty, D.J., McKeegan, D.E.F., 2015. Skin temperature reveals the intensity of acute stress. Physiology and Behavior 152: 225-230. https://doi.org/10.1016/j.physbeh.2015.09.032

Hoffmann, G., Herbut, P., Pinto, S., Heinicke, J., Kuhla, B., Amon, T., 2020. Animal-related, non-invasive indicators for determining heat stress in dairy cows. Biosystems Engineering 199: 83-96. https://doi.org/10.1016/j.biosystemseng.2019.10.017

Hovinen, M., Siivonen, J., Taponen, S., Hänninen, L., Pastell, M., Aisla, A.-M., Pyörälä, S., 2008. Detection of clinical mastitis with the help of a thermal camera. Journal of Dairy Science 91: 4592-4598. https://doi.org/10.3168/jds.2008-1218

Huffman, L., n.d. Bell laboratories and the development of electrical recording. Leopold Stokowski site. Available at: https://tinyurl.com/3hrd9jb5.

Huhtala, A., Suhonen, K., Mäkelä, P., Hakojärvi, M., Ahokas, J., 2007. Evaluation of instrumentation for cow positioning and tracking indoors. Biosystems Engineering 96: 399-405. https://doi.org/10.1016/j.biosystemseng.2006.11.013

Johnson, R.C., 2014. MEMS Mics taking over. Hot technologies: looking ahead to 2015. Available at: https://www.eetimes.com/mems-mics-taking-over/#.

Kaplan, E.D., Hegarty, C.J. (eds), 2006. Understanding GPS: principles and applications, 2nd ed. Artech House, Boston, MA, USA.

Keegan, K.G., Yonezawa, Y., Pai, P.F., Wilson, D.A., 2002. Accelerometer-based system for the detection of lameness in horses. Biomedical Sciences Instrumentation 38: 107-112.

Kjærgaard, M.B., Blunck, H., Godsk, T., Toftkjær, T., Christensen, D.L., Grønbæk, K., 2010. Indoor positioning using GPS revisited. In: Floréen, P., Krüger, A., Spasojevic, M. (eds) Pervasive computing. Proceedings of the 8th International Conference, Pervasive 2010, Helsinki, Finland, May 17-20, 2010. Lecture Notes in Computer Science (Including Subseries Lecture Notes in Artificial Intelligence and Lecture Notes in Bioinformatics) 6030 LNCS, pp. 38-56. https://doi.org/10.1007/978-3-642-12654-3_3

Krieger, S., Sattlecker, G., Kickinger, F., Auer, W., Drillich, M., Iwersen, M., 2018. Prediction of calving in dairy cows using a tail-mounted tri-axial accelerometer: a pilot study. Biosystems Engineering 173: 79-84. https://doi.org/https://doi.org/10.1016/j.biosystemseng.2017.11.010

Küntzel, A., Oertel, P., Trefz, P., Miekisch, W., Schubert, J.K., Köhler, H., Reinhold, P., 2018. Animal science meets agricultural practice: Preliminary results of an innovative technical approach for exhaled breath analysis in cattle under field conditions. Berliner Und Münchener Tierärztliche Wochenschrift 131: 417-443. https://doi.org/10.2376/0005-9366-17101

Lavanaya, P.G., Mallappa, S., 2019. Activity recognition from accelerometer data using symbolic data approach. In: Nagabhushan, P., Guru, D.S., Shekar, B.H., Sharath Kumar, Y.H. (eds.) Data analytics and learning. Lecture Notes in Networks and Systems. Springer Nature, Singapore, pp. 317-329. https://doi.org/10.1007/978-981-13-2514-4_27

Lee, S.M., Lee, D., Kim, Y.S., 2019. The quality management ecosystem for predictive maintenance in the Industry 4.0 era. International Journal of Quality Innovation 5: 4. https://doi.org/10.1186/s40887-019-0029-5

Lee, W.S., Lee, S.S., 2008. Piezoelectric microphone built on circular diaphragm. Sensors and Actuators A 144: 367-373. https://doi.org/10.1016/j.sna.2008.02.001

Lokhorst, K., 2018. Smart dairy farming. Van Hall Larenstein, Leeuwarden, the Netherlands. https://doi.org/10.31715/20181

Long, M.H., Rheuban, J.E., Berg, P., Zieman, J.C., 2012. A comparison and correction of light intensity loggers to photosynthetically active radiation sensors. Limnology and Oceanography: Methods 10: 416-424. https://doi.org/https://doi.org/10.4319/lom.2012.10.416

Luinge, H.J., Veltink, P.H., Baten, C.T.M., 1999. Estimation of orientation with gyroscopes and accelerometers. In: Proceedings of the First Joint BMES/EMBS Conference. 1999 IEEE Engineering in Medicine and Biology 21st Annual Conference and the 1999 Annual Fall Meeting of the Biomedical Engineering Society (Cat. N. 844 vols.2). https://doi.org/10.1109/IEMBS.1999.803999

Lüttenberg, H., Bartelheimer, C., Beveringen, D., 2018. Designing predictive maintenance for agricultural machines. In: Twenty-Sixth European Conference of Information Systems (ECIS2018). Portsmouth, UK, pp. 1-17.

Maddison, R., Ni Mhurchu, C., 2009. Global positioning system: a new opportunity in physical activity measurement. International Journal of Behavioral Nutrition and Physical Activity 6: 73. https://doi.org/10.1186/1479-5868-6-73

Maertens, W., Vangeyte, J., Baert, J., Jantuan, A., Mertens, K.C., De Campeneere, S., Pluk, A., Opsomer, G., Van Weyenberg, S., Van Nuffel, A., 2011. Development of a real time cow gait tracking and analysing tool to assess lameness using a pressure sensitive walkway: the GAITWISE system. Biosystems Engineering 110: 29-39. https://doi.org/https://doi.org/10.1016/j.biosystemseng.2011.06.003

Marshall, R.N., Harry, W.R., 1939. New microphone providing uniform directivity over an extended frequency range. Journal of the Acoustical Society of America 11: 164.

Martín Ocaña, S., Cañas Guerrero, I., González Requena, I., 2004. Thermographic survey of two rural buildings in Spain. Energy and Buildings 36: 515-523. https://doi.org/https://doi.org/10.1016/j.enbuild.2003.12.012

McDowell, G., 2009. How does a digital camera work. Technology explained. Available at: https://tinyurl.com/4d2wysmt.

Middleton, W.E.K., 1966. A history of the thermometer and its use in meteorology. John Hopkins Press, Baltimore, CA, USA.

Mortola, J.P., Kim, J., Lorzadeh, A., Leurer, C., 2015. Thermographic analysis of the radiant heat of chicken and duck eggs in relation to the embryo's oxygen consumption. Journal of Thermal Biology 48: 77-84. https://doi.org/10.1016/j.jtherbio.2015.01.001

Mottram, T., 2016. Animal board invited review: precision livestock farming for dairy cows with a focus on oestrus detection. Animal 10: 1575-1584. https://doi.org/10.1017/S1751731115002517

Muddle, J.R., 1984. Livestock weighing apparatus. Patent 4427083. Available at: https://patents.google.com/patent/US4427083.

Nääs, I.A., Garcia, R.G., Caldara, F.R., 2014. Infrared thermal image for assessing animal health and welfare. Journal of Animal Behaviour and Biometeorology 2: 66-72. https://doi.org/10.14269/2318-1265/jabb.v2n3p66-72

Nanda, M.B., Gouda, N., Rohan bidappa, A.C., Prajwal, S.V., 2020. Livestock monitoring using internet of things. In: Mahaboob Basha, S., Kumar, T.N.V.L.N., Raghava Reddy, P., Nanjesh, B.R. (eds.) Proceedings of the National Conference on Emerging Technologies in Engineering – 2020. Geethanjali Institute of Science and Technology, Nellore, India, p. 394.

Neethirajan, S., Tuteja, S.K., Huang, S.T., Kelton, D., 2017. Recent advancement in biosensors technology for animal and livestock health management. Biosensors and Bioelectronics 98: 398-407. https://doi.org/10.1016/j.bios.2017.07.015

Nichols, G. de la M., 1972. Electronic 'end of milking' detector and teat cup remover. New Zealand Journal of Agricultural Research 15: 639-642. https://doi.org/10.1080/00288233.1972.10430554

Nielen, M., Deluyker, H., Schukken, Y.H., Brand, A., 1992. Electrical conductivity of milk: measurement, modifiers, and meta analysis of mastitis detection performance. Journal of Dairy Science 75: 606-614. https://doi.org/10.3168/jds.S0022-0302(92)77798-4

O'Leary, N.W., Byrne, D.T., O'Connor, A.H., Shalloo, L., 2020. Invited review: cattle lameness detection with accelerometers. Journal of Dairy Science 103: 3895-3911. https://doi.org/10.3168/jds.2019-17123

Ordolff, D., 2001. Introduction of electronics into milking technology. Computers and Electronics in Agriculture 30: 125-149. https://doi.org/https://doi.org/10.1016/S0168-1699(00)00161-7

Pastell, M., Frondelius, L., Järvinen, M., Backman, J., 2018. Filtering methods to improve the accuracy of indoor positioning data for dairy cows. Biosystems Engineering 169: 22-31. https://doi.org/10.1016/j.biosystemseng.2018.01.008

Randle, H., Steenbergen, M., Roberts, K., Hemmings, A., 2017. The use of the technology in equitation science: A panacea or abductive science? Applied Animal Behaviour Science 190: 57-73. https://doi.org/10.1016/j.applanim.2017.02.017

Rosell-Polo, J.R., Auat Cheein, F., Gregorio, E., Andújar, D., Puigdomènech, L., Masip, J., Escolà, A., 2015. Advances in structured light sensors applications in precision agriculture and livestock farming. In: Sparks, D.L.B.T. (ed.) Advances in agronomy. Academic Press, London, UK, pp. 71-112. https://doi.org/https://doi.org/10.1016/bs.agron.2015.05.002

Rowe, E., Dawkins, M.S., Gebhardt-Henrich, S.G., 2019. A systematic review of precision livestock farming in the poultry sector: is technology focussed on improving bird welfare? Animals 9: 1-18. https://doi.org/10.3390/ani9090614

Schlecht, E., Hülsebusch, C., Mahler, F., Becker, K., 2004. The use of differentially corrected global positioning system to monitor activities of cattle at pasture. Applied Animal Behaviour Science 85: 185-202. https://doi.org/10.1016/j.applanim.2003.11.003

Seiko Suzuki, H., Shigeki Tsuchitani, M., Masayuki Miki, K., Masahiro Matsumoto, H., 1992. Capacitance type accelerometer. Patent 5095752. Available at: https://patents.google.com/patent/US5095752.

Sellier, N., Guettier, E., Staub, C., 2014. A review of methods to measure animal body temperature in precision farming. American Journal of Agricultural Science and Technology 2: 74-99. https://doi.org/10.7726/ajast.2014.1008

Shawn, 2020. What is a light sensor? Magazine 7 by AF Themes. Available at: https://tinyurl.com/44uu6m5f.

Sobel, D., 1995. Longitude. 4th Estate, London, UK.

Stafford, K.C., Collison, C.H., 1987. Manure pit temperatures and relative humidity of Pennsylvania high-rise poultry houses and their relationship to arthropod population development. Poultry Science 66: 1603-1611. https://doi.org/10.3382/ps.0661603

Stewart, M., Webster, J.R.R., Schaefer, A.L.L., Cook, N.J.J., Scott, S.L.L., 2005. Infrared thermography as a non-invasive tool to study animal welfare. Animal Welfare 14: 319-325.

United States Government, n.d. Official U.S. government information about the global positioning system (GPS) and related topics. Image Library. Available at: https://www.gps.gov/multimedia/images/.

Van den Berg, K., 1998. Milking system including a milk quantity meter. Patent 5792964. Available at: https://patents.google.com/patent/US5792964.

Venkatesh, V.C., Izman, S., 2008. Precision engineering. McGraw-Hill Professional, New Delhi, India.

Viscarolasaga, E., 2008. Local forms strum the chords of real music innovation. Journal of New England Technology, February 7, 2008. Available at: https://tinyurl.com/2jcx855w.

Voss, A.L., Fischer-Tenhagen, C., Bartel, A., Heuwieser, W., 2020. Sensitivity and specificity of a tail-activity measuring device for calving prediction in dairy cattle. Journal of Dairy Science 104: 3353-3363. https://doi.org/10.3168/jds.2020-19277

Watson, N.F., 1986. Accelerometer System. Patent 4601206. Available at: https://patents.google.com/patent/US4601206A.

Yoshida, S., 1984. Dynamic microphone. Patent 4427845. Available at: https://patents.google.com/patent/US4427845A/en.

Zehner, N., Umstätter, C., Niederhauser, J.J., Schick, M., 2017. System specification and validation of a noseband pressure sensor for measurement of ruminating and eating behavior in stable-fed cows. Computers and Electronics in Agriculture 136: 31-41. https://doi.org/10.1016/j.compag.2017.02.021

9. Data science applications in farms animals and companion animals

G. Hofstra and M.T.J. Terlien*

HAS University of Applied Sciences, P.O. Box 90108, 5200 MA 's Hertogenbosch, the Netherlands; m.terlien@has.nl

Highlights

- Data science is about answering questions by combining and analysing digital sources, which contain data.
- Data science has close relations to subjects like big data, data mining and machine learning.
- The data science workflow exists of business understanding, data understanding, data preparation, modelling, evaluation and deployment.
- Most business objectives can be translated into one of the following data science problems: classification problem, clustering problem, regression problem.
- Classification is a supervised learning method that categorises unlabelled observations based on a set of labelled observations.
- Regression is a supervised learning method that uses labelled observations to quantify the relationship between a dependent variable (the label) and one or more independent variables.
- Clustering is an unsupervised learning method that groups unlabelled observations into clusters with similar characteristics.
- There are several user interface based tools and scripting languages available to solve data science problems.
- Deep learning and vision technology hold great promise for the livestock sector.

G. Hofstra and M.T.J. Terlien

9.1 Why use data science (data-driven decisions)

To answer the question of why we should use data science, we should first take a look at what data science is. Data science in its definition is a concept of unifying statistics, data analysis with their related methods and the eventual results. It involves principles, processes, and techniques for understanding phenomena via the (automated) analysis of data with as goal being able to support decision making (Hayashi, 1998; Provost and Fawcett, 2013).

Put differently, data science is about answering questions by combining and analysing digital sources, which contain data. It is important to realise that data science is as much about science as it is about data (Leek, 2013). It is easy to discover correlations in a data set; there are always correlations to be found for a host of reasons if you simply collect enough data. Being able to understand whether these correlations are in any way relevant for a specific issue is a lot more challenging.

Data science has close relations to subjects like big data, data mining and machine learning.

- ▶ Big data can be defined as high-volume, high velocity and/or high variety-information assets that demand cost-effective, innovative forms of information processing that enable enhanced insight, decision making and process automation (Gartner, 2020). Put simply, big data is about large and often more complex data sets. These data sets are so voluminous that traditional data processing software cannot manage them anymore. But these huge sets of data can be used to address business problems we would not have been able to tackle without them.
- ▶ Data mining is the process of discovering patterns and relationships in data. The field combines tools from statistics and artificial intelligence (such as machine learning) with database management to analyse data sets (Clifton, 2019).
- ▶ Machine learning is the study of computer algorithms that improve automatically through experience and it is seen as a subset of artificial intelligence. Machine learning algorithms build a mathematical model based on sample data, known as 'training data', to make predictions or decisions (Mitchell, 1997).

9. Data science applications in farms animals and companion animals

Now we know what data science is, why, then, should we apply it to our farming practices or use it in companion animals? Let's take the pig disease enzootic pneumonia as an example. Enzootic pneumonia caused by *Mycoplasma hyopneumoniae* has a significant impact on the performance in pig stables. It causes a decrease in feed efficiency and mean daily weight gain which are both correlated to the severity of the pneumonia. Sensors can generate a continuous stream of data which in turn gives us the chance to analyse current but also past events. Based on past events, we might be able to give predictions of future events and thus improve our decision making. For instance, by using data science on sensor data of previous *Mycoplasma* outbreaks in a pig stable we might be able to find patterns in the onset of the past outbreaks (e.g. 5% increase in coughing + 10% decrease of feed intake) which could give us an early warning when similar patterns occur. Not only could the recognition of patterns be an aid in preventing or recognising disease but also in optimising business performance.

9.2 Data science workflow

Several phases can be distinguished in the data science workflow (Figure 9.1).

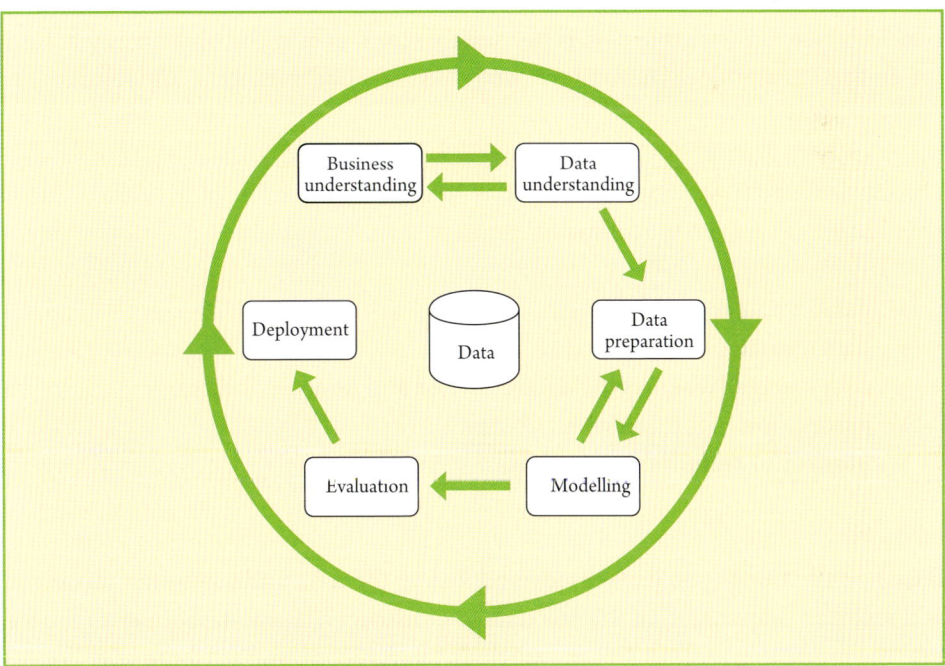

Figure 9.1. Data science workflow (Chapman *et al.*, 2000).

9.2.1 Phases in data science workflow

Business understanding

This first phase focuses on the goals of the project and thus understanding the objectives and requirements for the business. In this phase, one should also identify which data is already available. Once the objectives are clear they can be transformed into a data science problem and a plan can be drafted to achieve the objectives. Most objectives can be translated into one of the following data science problems:

- classification problem;
- clustering problem;
- regression problem.

(See Section 9.3 for the explanation of these data science problems)

Data understanding

Data understanding starts with the initial data collection. After this, it is all-important to explore your data and become more familiar with the data. By simply exploring the data you can for instance identify issues with data quality or have some first insights about how to reach your objectives. During the process of data collection, you may conclude that the available data is not suitable for reaching the objectives you have set. In that case, you will have to return to business understanding and reformulate your objectives.

During data understanding, a list must be compiled of all relevant datasets. This list should contain at least the following items:

- dataset name;
- data format;
- data storage (where is the dataset stored and its accessibility);
- description of the dataset attributes;
- number of records of each dataset.

Data preparation

Based on the things you have discovered during the data understanding phase you will start the preparation of the raw data to construct a final dataset that can be fed into the modelling tool(s). This may include table, record, and attribute selection, as well as transformation and cleaning of data for modelling tools.

Be aware! It is estimated that approximately 80% of the time of a data science project is spent on data preparation.

Modelling

In the modelling phase, several techniques might turn out to be suitable for the same problem and thus it is important to explore these possibilities. However, since different techniques may have different requirements on the form of the data it might often be necessary to return to the data preparation phase and adjust the dataset accordingly. The selection of a proper algorithm is a matter of experience.

Evaluation

In this phase, you will be evaluating the steps taken up to now to assure that the model(s) properly achieve(s) the business objectives. Are there any business issues being overlooked? In the end, a decision on the use of the data mining results should be reached.

After running, for instance, a classification or regression to find the model, you have to assess the quality of the model. For supervised models, the quality of the model can be assessed by determining how well your model predicts the label (class or value) of new inputs that have not been used to derive the model. This process is called model testing. Testing results are also discussed with domain experts to evaluate whether it meets the business objectives.

In clustering problems, the algorithm is used to cluster the input in several clusters. The quality of a clustering is best evaluated by contacting domains experts to see whether the algorithm resulted in meaningful clusters and business objectives are met.

Deployment

Finally, the knowledge gained by the model will have to be organised and presented in such a way that the business can use it. Depending on the project requirements this can range from reports being generated to repeatable data mining processes being implemented across a business.

9.2.2 Good practices

To be able to ensure the science part within data science there are some good practices you might want to abide by:

- Collect independent measurements and when possible, record them machine-readable (e.g. numbers instead of words).
- Take the limitations of your measurements and other data used into account (How precise are your measurements? How reliable/relevant is the data?).
- Use sound scientific reasoning (Use sound logic, sound math, sound statistics, sound code for analysis. Make minimal assumptions and question them!).
- Use FAIR data (Findable, Accessible, Interoperable, Reusable).

9.3 Overview of data science techniques including examples

9.3.1 Machine learning

Models are developed to make predictions. In traditional modelling, the investigator develops a model (a set of rules) to map input (observations) to output (predictions). In machine learning, a computer calculates the model to map the input to output (Figure 9.2). Therefore, machine learning is described as 'The field of machine learning is concerned with the question of how to construct computer programs that automatically improve with experience' (Mitchell, 1997).

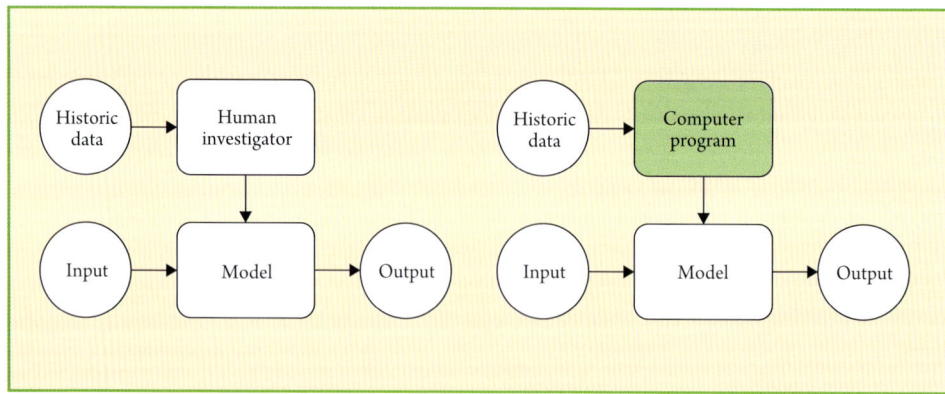

Figure 9.2. Machine learning paradigm.

In machine learning the individual measurable property of an observation (e.g. weight, age, length, colour) is called a feature or attribute. There are four different measurements scales for features (nominal, ordinal, interval and ratio).

Machine learning algorithms (algorithms to build the model) can be classified into supervised and unsupervised learning algorithms (Figure 9.3). Other machine learning algorithms such as reinforced algorithms are beyond the scope of this book.

Supervised learning

In supervised learning, we develop a predictive model using observations (input features) with known outcomes (labels). This model is then used to predict the label of unlabelled observations. Supervised learning starts with collecting a set of labelled observations (labelled dataset). This labelled dataset is split into a training dataset, a validation dataset and a test dataset. The training data is used to train the supervised learning algorithm (model) to accurately predict the known labels. Then the validation dataset is used to tune the parameters that control the learning process (the hyperparameters). An example of a hyperparameter is the depth of a decision tree (see Section 9.3.3). Finally, the test dataset is used to evaluate the performance of the model on new previously unseen input. The validated and tested model is ready for use on an unlabelled dataset. This workflow for building a supervised learning model is shown in Figure 9.4.

When the model performs well on the training dataset but not on previously unseen input (the validation dataset) the model is overfitted. A possible solution to overfitting is reducing the model complexity, for example by using fewer features or reducing the

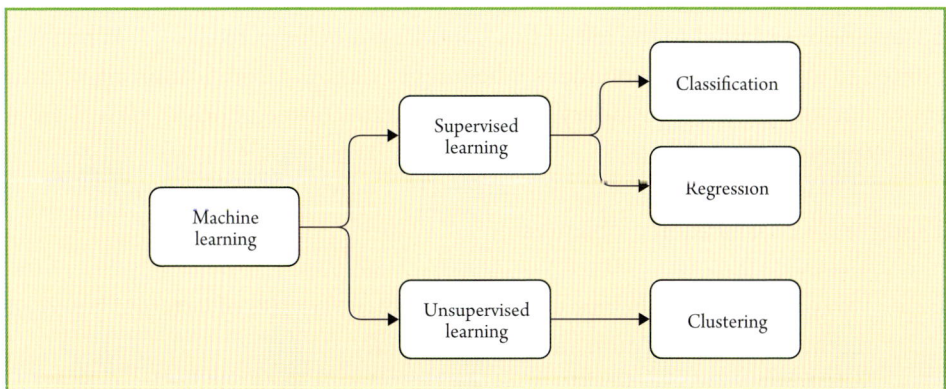

Figure 9.3. Classification of machine learning algorithms.

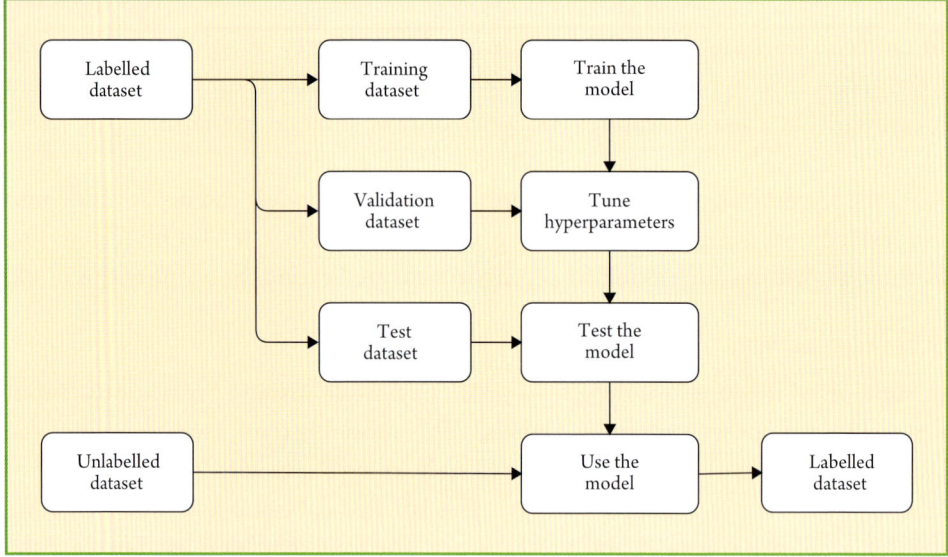

Figure 9.4. Workflow supervised learning.

depth of the decision tree (see Section 9.3.3). A model is underfitted when it does not perform well on the training set. The best remedy against underfitting is trying a different machine learning algorithm or increasing the model complexity. A good model is neither underfitted nor overfitted.

When the labels to predict are classes, we are dealing with a classification problem. The concept of classification is illustrated in Figure 9.5. This figure shows a plot of a dataset with *input_feature_1* on the x-axis and *input_feature_2* on the y-axis. The output label is shown as a cross (for observations of *class_1*) or a square (for observations of *class_2*). The line represents the decision boundary. When an observation is plotted to the left of this boundary, it belongs to *class_1*. When it is plotted to the right of this boundary, it belongs to *class_2*. Classification algorithms try to find the decision boundary that best separates observations of *class_1* from observations of *class_2*.

Examples of classification problems are:

- Does a cow have subclinical mastitis or not at day 100 of lactation based on the somatic cell count and the milk yield?
- What size eggs (Small, Medium, or Large) will Barnevelder layer hens produce based on age and feed intake?

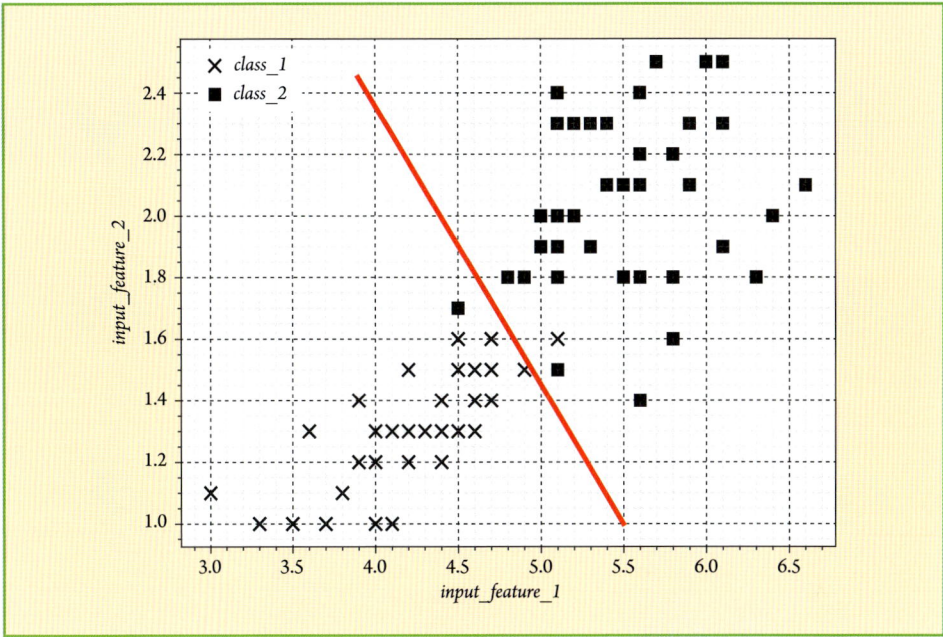

Figure 9.5. Illustration of a classification problem.

In the first example, the somatic cell count and milk yield loss are features. Whether a cow has subclinical mastitis or not is the label. In the second example, age and feed intake are features and the size of the eggs is the label.

When the labels to predict are numerical values, we are dealing with a regression problem.

The concept of regression is illustrated in Figure 9.6. This figure shows a plot of a dataset with the input feature on the x-axis and the output label on the y-axis. The line is the linear relation between the input feature and output label and is used to predict the output label for unseen input features. For example: When the input feature has a value of 34 the predicted output label is 50.

Examples of regression problems are:

- How much milk yield loss is there based on somatic cell count in cows with subclinical mastitis?
- How many eggs (in grams) will Barnevelder layer hens produce per week based on their age?

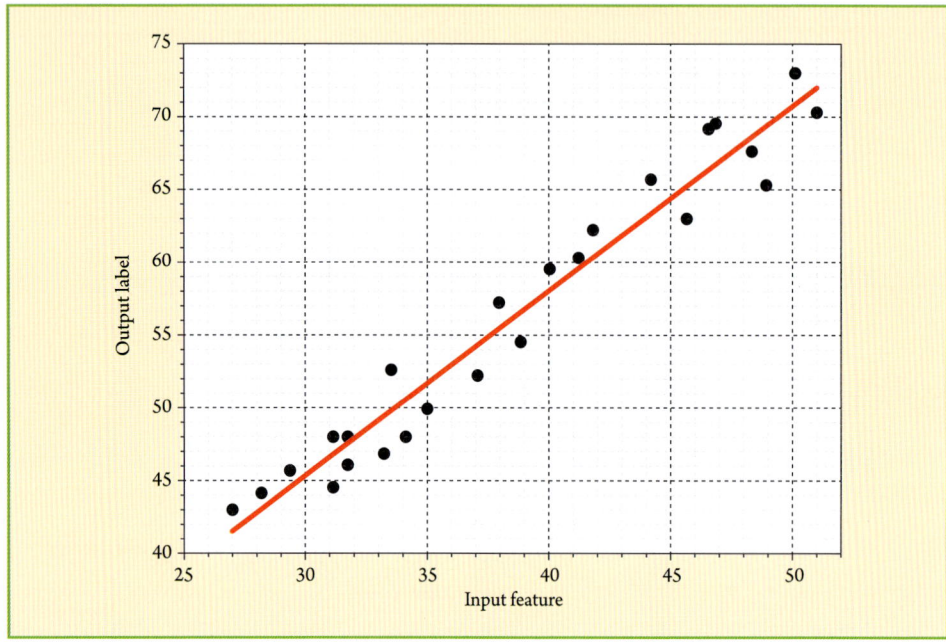

Figure 9.6. Illustration of a regression problem.

In regression problems, the features are the independent variables while the label is the dependent variable. In the first example, the somatic cell count is the independent variable, and the milk yield loss is the dependent variable. In the second example, age is the independent variable, and the total weight of the eggs is the dependent variable.

Unsupervised learning

Unsupervised learning models are developed based on input data only to group and interpret the input. The computer is used to cluster unlabelled input data into groups with similar characteristics. An example of an unsupervised learning method is clustering (Figure 9.7). In this figure, four clusters are identified based on two input features, shown on the x-axis and y-axis. Observations in the same cluster are more similar than observations in different clusters. In clustering algorithms, the number of clusters is not known beforehand.

An example of a clustering problem is:

▶ Can we cluster horse enthusiasts based on the way they search and find out information, their emotional involvement with horses, and their attitude, their knowledge, and daily practices regarding the welfare of horses (Visser et al., 2012)

9. Data science applications in farms animals and companion animals

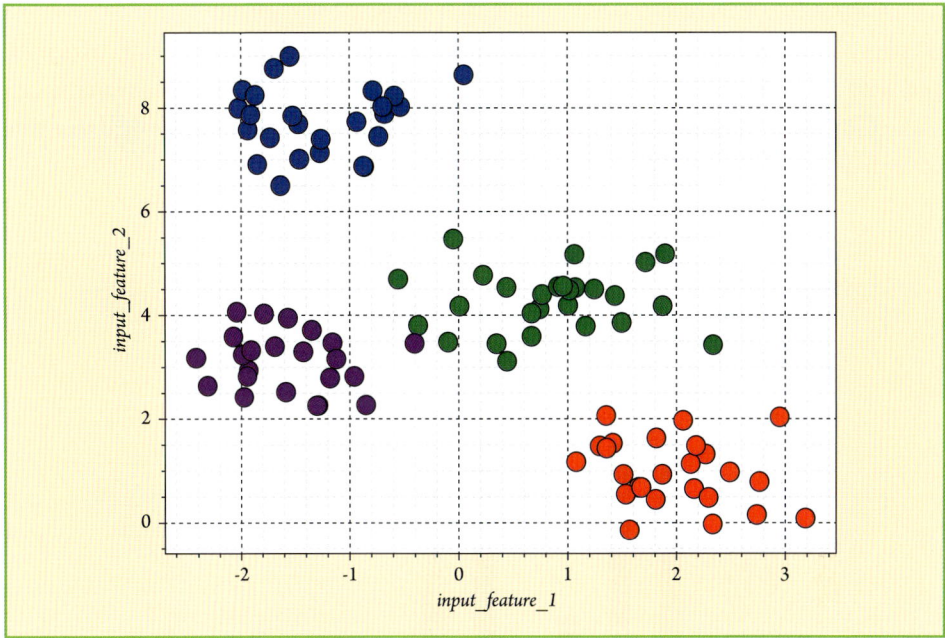

Figure 9.7. Illustration of the clustering problem.

9.3.2 Nearest neighbour and *k*-nearest neighbour classification

Nearest neighbour and *k*-nearest neighbour are supervised classification methods. Both methods store the entire training set. For any new observation, it finds the *k* most similar observations in the training set, known as neighbours, where *k* represents the number of most similar neighbours taken into consideration. These *k* neighbours are then used to label the new observation according to some prediction scheme. One possibility is to assign the label that is most common in the neighbourhood.

To illustrate how to use these methods, we use a fictional dataset of cows with subclinical mastitis. Our example dataset consists of 120 cows, 60 cows have subclinical mastitis and 60 cows do not have mastitis. For each cow, the somatic cell count, milk yield at day 100 of the lactation and if she has subclinical mastitis or not is known.

The dataset is randomly split into a training dataset and a test dataset. The training dataset is used to tune the value of *k*. The test dataset is set apart and is not used for exploratory analysis and model building. It is left until we have chosen our final model to assess how well this model performs on new data.

As part of the exploratory analysis, we make histograms of the number of cows with and without subclinical mastitis per milk yield bin (Figure 9.8) and per somatic cell count bin (Figure 9.9). These histograms show that cows with subclinical mastitis have in general a lower milk yield and a higher somatic cell count.

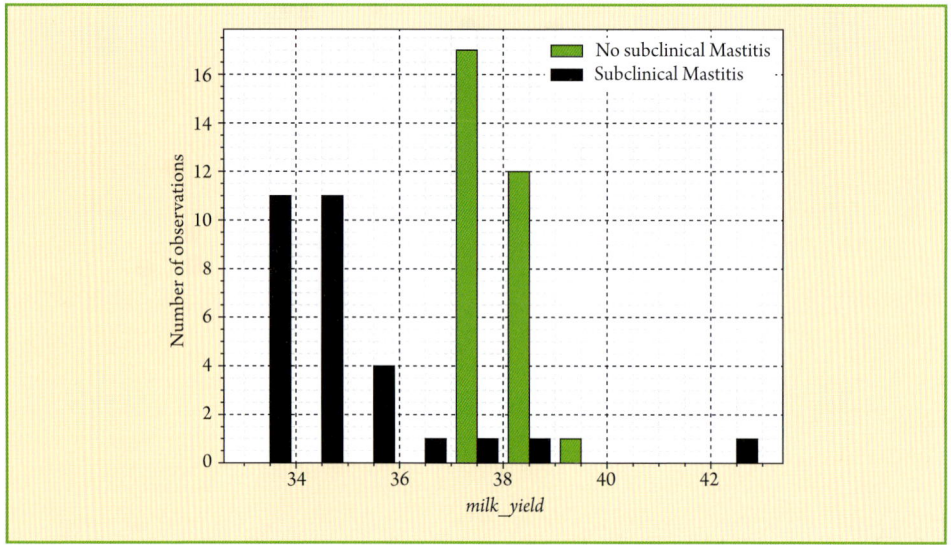

Figure 9.8. Histogram of milk yield (litres) at day 100 of the lactation.

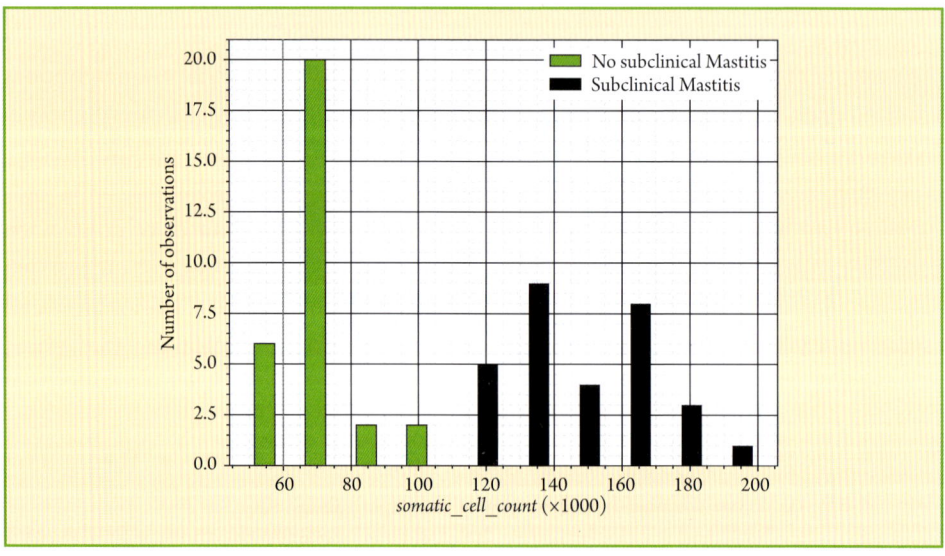

Figure 9.9. Histogram of somatic cell count.

Outlier detection is another important step in the exploratory analysis. An outlier is an observation that differs significantly from other observations. It may be the result of variability in your data or a measurement error. Outliers are sometimes excluded from further analysis as they can seriously influence the outcome. Possible outliers can be identified with a box plot. From the boxplots in Figure 9.10 and 9.11, it can be concluded that there are some outliers in the dataset (the circles in the boxplot).

To explain the concept of nearest neighbour classification four observations of cows with subclinical mastitis and three observations of cows without subclinical mastitis are plotted in a graph with milk yield on the x-axis and somatic cell count on the y-axis. A new cow with a known milk yield and known somatic cell count is also plotted in the graph (Figure 9.12).

When the somatic cell count and milk yield of the new cow are similar to the somatic cell count and milk yield of the cow with subclinical mastitis, the new cow is most likely to have subclinical mastitis. When the somatic cell count and milk yield are similar to the somatic cell count and milk yield of a cow without mastitis the new cow most likely does not have subclinical mastitis.

To quantify 'similar to' we must define a distance metric. A simple distance metric is the Euclidian distance. The Euclidian distance is calculated as the square root of the sum of the squared differences between the features of the labelled observation and the features of the unlabelled observation.

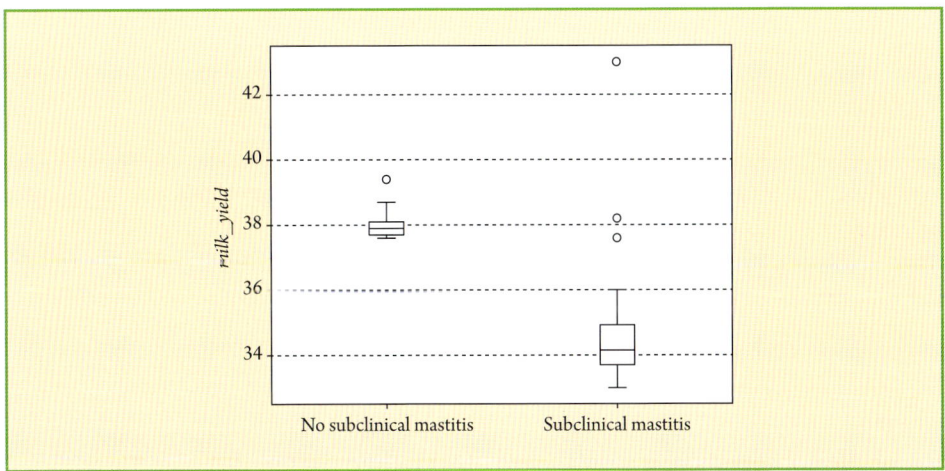

Figure 9.10. Boxplot milk yield.

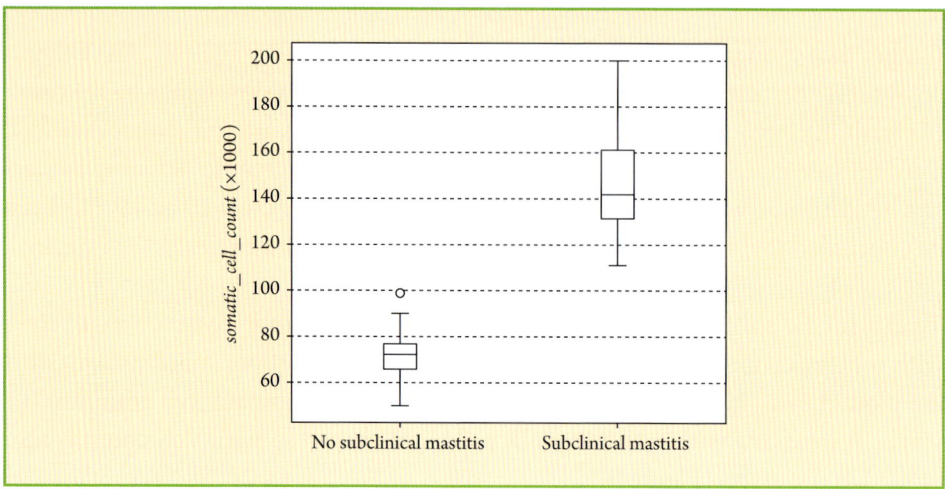

Figure 9.11. Boxplot somatic cell count.

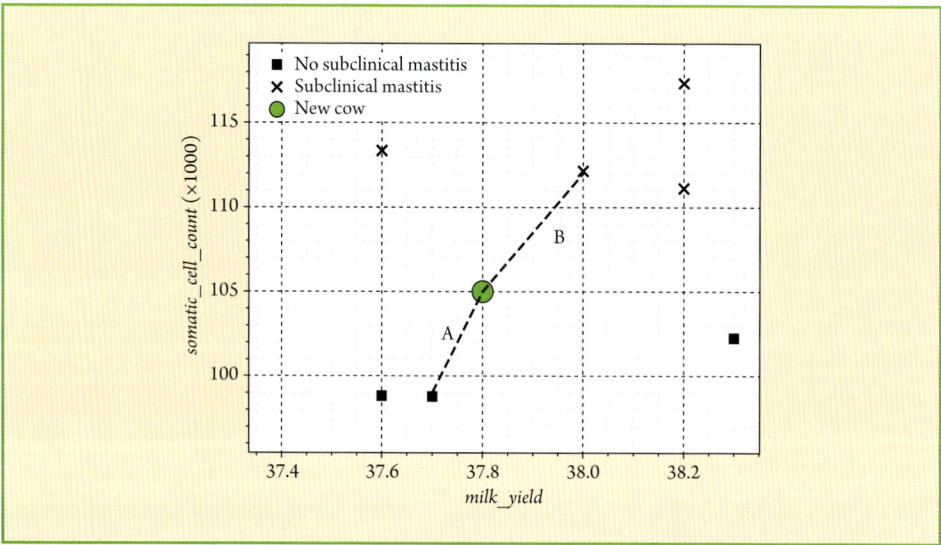

Figure 9.12. Plot of cows with and without subclinical mastitis as a function of milk yield and somatic cell count.

$$distance\ (p, q) = \sqrt{\sum_{i=1}^{n}(p_i - q_i)^2}$$

where:
p = labelled observation; q = unlabelled observation; n = number of features.

In the case of two features, the distance measure reduces to the Pythagorean theorem:

$$distance\ (p, q) = \sqrt{(p_1 - q_1)^2 + (p_2 - q_2)^2}$$

The smaller the Euclidian distance the closer the observations are to each other.

Other distance metrics are for example Manhattan distance and Mahalanobis distance. The choice of distance measure should be made, based on an analysis of the nature of the features and an analysis of the performance of the resulting model.

To illustrate the classification process, we use the Euclidian distance as the distance metric. The Euclidean distance A (to the closest cow without subclinical mastitis) is:

$$\sqrt{(37.8 - 37.7)^2 + (105{,}000 - 99{,}000)^2} = 6{,}000$$

The Euclidean distance B (to the closest cow with subclinical mastitis) is:

$$\sqrt{(37.8 - 38.0)^2 + (105{,}000 - 112{,}000)^2} = 7{,}000$$

Because the Euclidean distance to the cow without mastitis (6,000) is smaller than the Euclidean distance to the cow with subclinical mastitis (7,000), the new cow is more similar to the cow without mastitis and therefore we classify or label her as a cow without mastitis.

In our example, the somatic cell count varies between 50,000 and 200,000 while the milk yield varies between 33 and 43 litres. In classification problems, the feature with the largest scale range (in this example the somatic cell count) has the largest influence on the result. Due to the difference in scale range, the label of an unknown observation is entirely determined by the difference in somatic cell count. The difference in milk yield does not play a role at all.

To make your result independent of the scale ranges of your input data the features should be normalised. Normalisation is the process of giving your input features a common scale (often a scale of 0 to 1). The min-max normaliser linearly rescales every feature to the $[0,1]$ interval:

$$x_{new} = \frac{x_{old} - \min(x_{old})}{[\max(x_{old}) - \min(x_{old})]}$$

We use the min-max normaliser to normalise our input features. Table 9.1 and 9.2 show the effect of normalisation. Both features have the same minimum and maximum value, and all values are between 0 and 1.

Now that we have chosen a metric to quantify similarity and normalised our features, we can build the model. We run the nearest neighbour classification for different values of k to find the value of k that gives the model with the highest accuracy. The accuracy of the model is defined as the number of correctly labelled observations divided by the total number of observations. To calculate the accuracy of the model a confusion matrix is created for each value of k. In a confusion matrix, the labels of the observations in the validation dataset are compared with the labels predicted by the model. The columns of the matrix are the predicted labels while the rows of the matrix are the true labels in the validation dataset.

Table 9.3 shows the confusion matrix for $k=1$. All cows with subclinical mastitis got the label 'Subclinical mastitis'. From the cows that do not have subclinical mastitis 3 of the 24 cows got the wrong label 'Subclinical mastitis', 21 of 24 cows got the correct label 'No subclinical mastitis'.

Table 9.1. Milk yield before and after normalisation.

	Before	After
Minimum value	33.0	0
Mean value	36.6	0.34
Maximum value	43.0	1
Range	10.0	1

Table 9.2. Somatic cell count before and after normalisation.

	Before	After
Minimum value	51,386	0
Mean value	106,459	0.41
Maximum value	198,473	1
Range	147,087	1

Table 9.3. Confusion matrix for nearest neighbour classification subclinical mastitis ($k=1$).

	Predicted label	
	No subclinical mastitis	Subclinical mastitis
No subclinical mastitis	21 ✓	3 ✗
Subclinical mastitis	0 ✓	24 ✓

In our example 45 of out 48 cows are correctly labelled resulting in an accuracy of 45/48 = 0.94. An accuracy of 0.94 means that in 94% of the cases the model correctly predicts the label of an unlabelled observation.

The results of running a k-nearest neighbour classification of our example dataset with different values for k is shown in Table 9.4.

With a value for k of 3 (assign the label based on the three closest observations) the accuracy of the model improves from 0.94 to 0.96 meaning that 96% of new observations are labelled correctly. Using a higher number of k does not lead to higher accuracy of the model.

The next step is to run the model with $k=3$ on the test dataset to get an unbiased estimate of its performance. Finally, a new model is built on the entire dataset (both training and test data) and that is the model that is used in production.

Table 9.4. Accuracy of k-nearest neighbour classification as a function of k.

Value for k	Accuracy
1	0.94
3	0.96
5	0.94
7	0.94

9.3.3 Decision tree classification

Decision tree classification is another classification algorithm. When most features are non-numeric a decision tree is a more suitable method than nearest neighbour or k-nearest neighbour. To illustrate how decision trees are constructed, we are going to use a fictional dataset of cows with and without clinical mastitis. Our training dataset consists of six cows with clinical mastitis and four cows without clinical mastitis. For each cow is registered whether she has flakes in milk, udder swelling and a warm udder. Flakes in milk, udder swelling, and a warm udder are an indication that a cow has clinical mastitis. The training dataset is shown in Table 9.5.

A decision tree is a tree-like model of questions, starting with a question (root) at the top (Figure 9.13). Each node represents a question and its branches represent the answer to that question. Finally, in the leaves of the tree, a classification is made. The objective of a decision tree is to find the sequence of questions that best predicts the label of previously unseen observations. Building a decision tree starts with finding the root question.

For our example dataset we have three possible questions to start with (one for each feature):

- ▶ Does a cow have flakes in its milk or not?
- ▶ Does a cow have udder swelling or not?
- ▶ Does a cow have a warm udder or not?

Table 9.5. Fictional example dataset of cows with and without clinical mastitis.

Cow	Flakes in milk	Udder swelling	Warm udder	Mastitis
1	true	true	true	true
2	true	true	true	true
3	true	false	true	true
4	false	true	true	false
5	true	true	true	true
6	true	true	true	true
7	true	false	false	false
8	true	false	true	false
9	true	true	true	true
10	false	false	true	false

9. Data science applications in farms animals and companion animals

For each question, there are two answers (True or False). All cows with a positive (True) answer go into one group. All cows with a negative (False) answer go into another group. To find the question (feature) that best separates cows with mastitis from cows without mastitis, we must quantify the homogeneity of the resulting groups. The more homogeneous the groups, the better the question separates cows with mastitis from cows without mastitis. The question that results in the most homogeneous (less impure) groups is the question to start with at the root of the decision tree.

One way to measure the homogeneity of a group is the Gini impurity. The Gini impurity has a value between 0 and 0.5. A group with an impurity of 0 is a homogeneous group. The closer the impurity is to 0.5, the more heterogeneous the group. The formula for the Gini impurity is given in the following equation,

$$Gini = 1 - \sum_{i=1}^{C}(p_i)^2$$

where C is the number of classes and p_i the probability of picking an observation of class i.

To find the impurity for a group, we first calculate the probability (p) for a cow with mastitis by dividing the number of cows with mastitis by the total number of cows in the group. Next, we calculate the probability for a cow without mastitis as $1 - p$. Finally, we square the two probabilities, sum the two squared probabilities and subtract the result from 1. We repeat this calculation for each group.

To compensate for the fact that the True and the False group usually do not contain the same number of observations, we take the weighted average of the group impurity to calculate the total Gini impurity for each question.

Tables 9.6, 9.7 and 9.8 give the impurities for the three questions.

The question 'Does the cow have flakes in milk' results in the lowest impurity, so this is the root question of our decision tree.

Table 9.6. Gini impurity for the question 'Does the cow have flakes in milk?'

Flakes in milk	True (1)		False (0)	
	Mastitis	No mastitis	Mastitis	No mastitis
	6 cows	2 cows	0 cows	2 cows
Probability	0.75	0.25	0	1
Gini impurity group	$1 - (0.75*0.75 + 0.25*0.25) = 0.375$		$1 - (0*0 + 1*1) = 0$	
Gini impurity feature	$(8/10) * 0.375 + (2/10) * 0 = 0.30$			

Table 9.7. Gini impurity for the question 'Does the cow have udder swelling?'

Udder swelling	True (1)		False (0)	
	Mastitis	No mastitis	Mastitis	No mastitis
	5 cows	1 cow	1 cow	3 cows
Probability	0.84	0.16	0.25	0.75
Gini impurity group	$1 - (0.84*0.84 + 0.16*0.16) = 0.27$		$1 - (0.25*0.25 + 0.75*0.75) = 0.38$	
Gini impurity feature	$(6/10) * 0.27 + (4/10) * 0.38 = 0.31$			

Table 9.8. Gini impurity for the question 'Does the cow have a warm udder?'

Warm udder	True (1)		False (0)	
	Mastitis	No mastitis	Mastitis	No mastitis
	6 cows	3 cows	0 cows	1 cow
Probability	0.67	0.33	0	1
Gini impurity group	$1 - (0.67*0.67 + 0.33*0.33) = 0.44$		$1 - (0*0 + 1*1) = 0$	
Gini impurity feature	$(9/10) * 0.44 + (1/10) * 0 = 0.40$			

In the next step, we repeat this process for the groups with an impurity larger than 0. We have two possible questions left:

- Does the cow have udder swelling?
- Does the cow have a warm udder?

The question 'Does the cow have udder swelling?' results in the lowest impurity, so this feature is the second question in our decision tree and 'Does the cow have a warm udder?' becomes the last question in our decision tree.

The final decision tree for our example dataset is shown in Figure 9.13.

For each question, we can calculate the number of correctly classified cows as a measure of the accuracy of the model (Table 9.9).

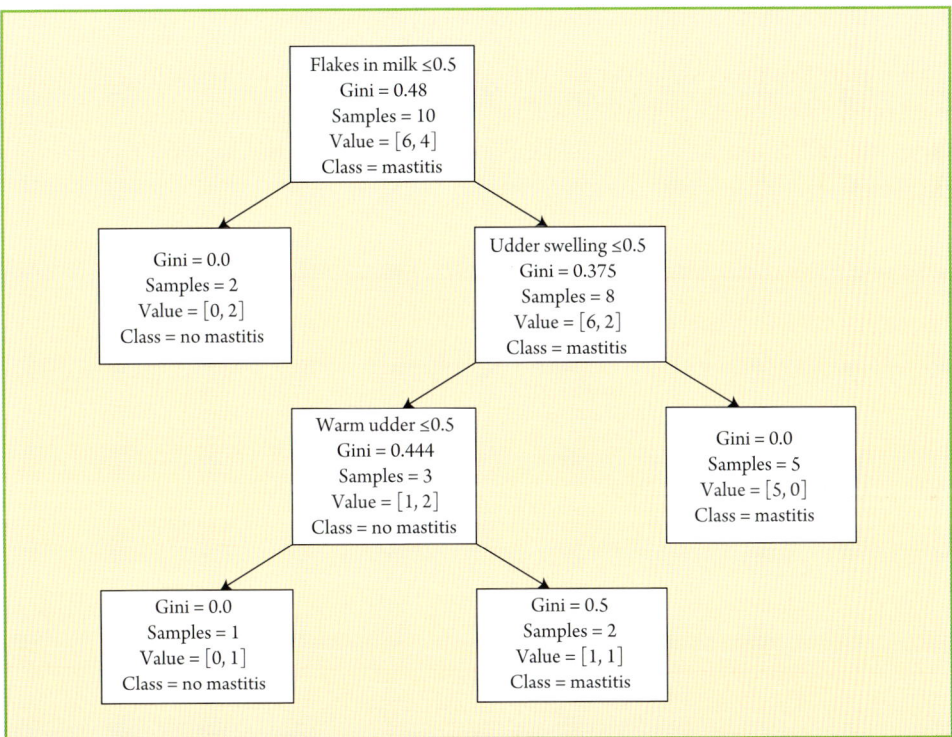

Figure 9.13. Decision tree for classifying cows with and without clinical mastitis.

Table 9.9. Model accuracy a function of decision tree question.

Question	Number of cows accurately predicted	Accuracy
Does a cow have flakes in milk?	8	0.8
Does a cow have udder swelling?	9	0.9
Does a cow have a warm udder?	9	0.9

When adding a question does not lead to a significant increase in model accuracy it should not be added because this can lead to overfitting. In this example, the last question does not lead to an increase in accuracy and should be left out of the decision tree. The validation dataset should be used to find the optimal number of questions (the depth of the decision tree).

It is also possible to construct a decision tree for continuous features. In the nodes is tested if the value of a feature exceeds a threshold or not. To find this threshold, we also use the Gini impurity. For each feature value, the Gini impurity is calculated and the value that results in the lowest Gini impurity is selected as the threshold.

To illustrate this process, we use the example dataset of cows with subclinical mastitis used for the nearest neighbour classification. This dataset has two numerical features (somatic cell count and milk yield). Figure 9.14 shows the decision tree for this dataset.

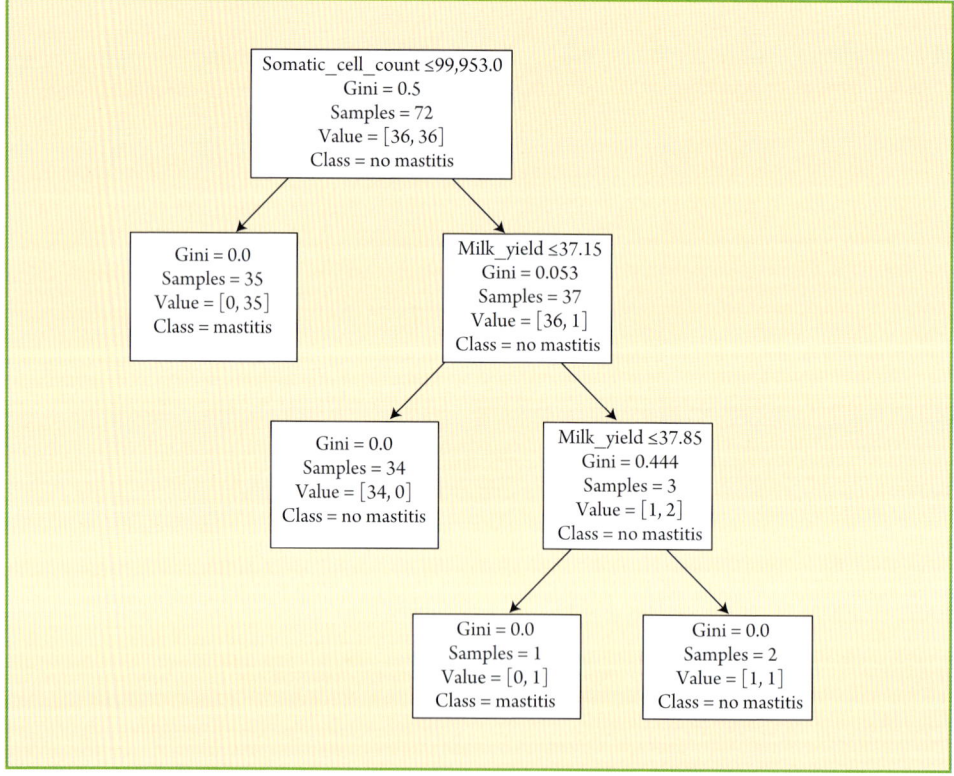

Figure 9.14. Decision tree for classifying cows with and without subclinical mastitis.

In this case, using only the root question the model has already an accuracy of 0.99 (71 out of 72 cows are labelled correctly). The two additional questions increase that accuracy to 1.00 at the cost of complexity and the risk of overfitting.

9.3.4 Simple linear regression

Regression is a supervised classification method used to predict numerical output. In simple linear regression problems, the numerical output (dependent variable) depends on only one independent variable and the relation between dependent and independent variable is linear.

To explain simple linear regression, we are going to use a fictional example dataset to predict the average egg weight (in grams) a Barnevelder layer hen will produce based on her age. In our example, the age of the Barnevelder layer hen is the independent variable and the egg weight is the dependent variable. Since we are dealing with only one independent variable it is a simple regression problem. In the case of more than one independent variable, we would have a multiple regression problem. Multiple regression is beyond the scope of this book.

Our example dataset consists of 50 hens. The dataset is split into a training dataset with 25 hens and a test dataset with 25 hens. For each hen, the age (in weeks) and the average egg weight (in grams) are known. The training dataset is plotted to get a better understanding of the relation between the age (independent variable) and the egg weight (dependent variable).

The plot in Figure 9.15 shows a linear relationship between age and egg weight. The egg weight increases in a linear way when the age of the hen increases. This means that we can use simple linear regression to describe the relationship between age and egg weight. Simple linear regression is about finding the line that best predicts the value of the dependent variable (egg weight) based on the value of the independent variable (age). This line is called the regression line and is described by the following equation:

$$Y = b + a \times X$$

where:
Y = dependent variable to predict
X = independent variable
a = slope of the line
b = intercept (value where line cuts y-axis)

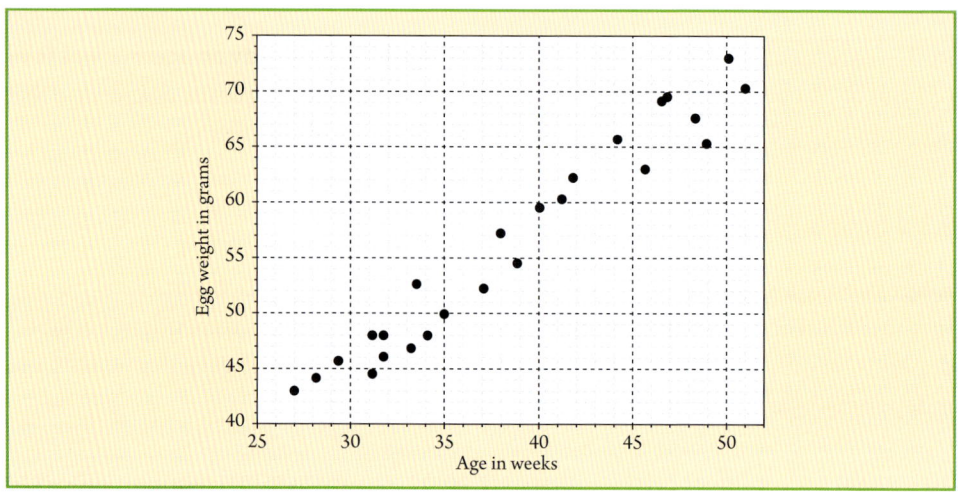

Figure 9.15. Plot of the relation between age and egg weight.

The ordinary least squares method is used to find the values for slope and intercept. The ordinary least squares method minimises the sum of squared residuals. The residual is defined as the difference between the observed and predicted value (Figure 9.16). The sum of the squared residuals (SSR) is defined as:

$$SSR = \sum_{i=1}^{n} (y_{observed} - y_{predicted})^2$$

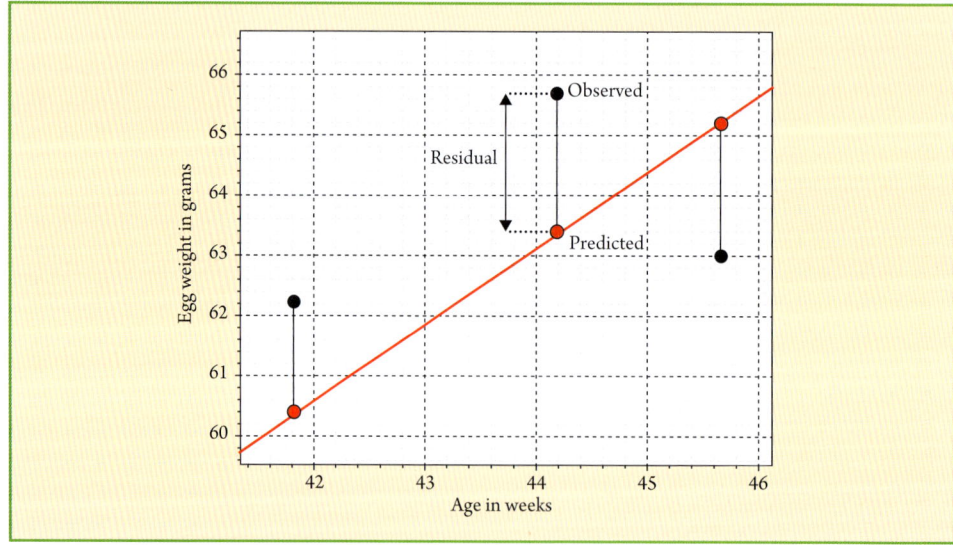

Figure 9.16. Calculation of residual.

Using the ordinary least squares method, we find the following regression line (model) for our training dataset. The line has a slope of 1.3 and an intercept of 7.3:

Egg weight = 7.3 + 1.3 × Age

A slope of 1.3 means that with every additional week the weight of the egg increases by 1.3 grams. The training dataset and the regression line are shown in Figure 9.17.

We run the model on the test dataset to evaluate its performance (Figure 9.18). The Root Mean Squared Error (RMSE) can be used to quantify the performance. It is a measure of the average distance between the value on the regression line predicted by our model and the actual value in the test dataset. It is calculated with the following formula:

$$Root\ Mean\ Squared\ Error = \sqrt{\sum_{i=1}^{n}(\hat{y}_i - y_i)^2}$$

where $\hat{y}_i$ is the value predicted by the model and y_i is the actual value in the test dataset.

For the test dataset, we find an RSME of 4.2. The smaller the RSME the better the model predicts the egg weight.

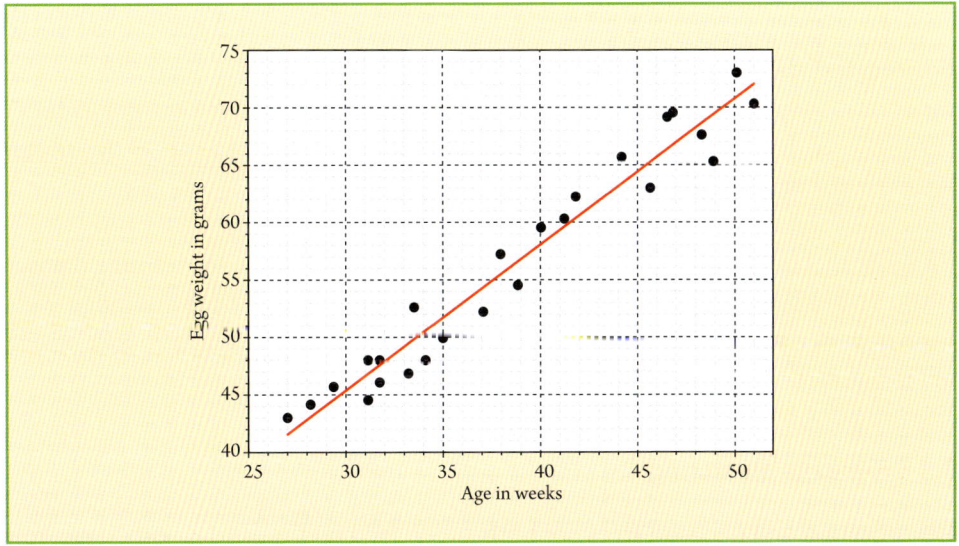

Figure 9.17. Plot of the training dataset and the regression line (model).

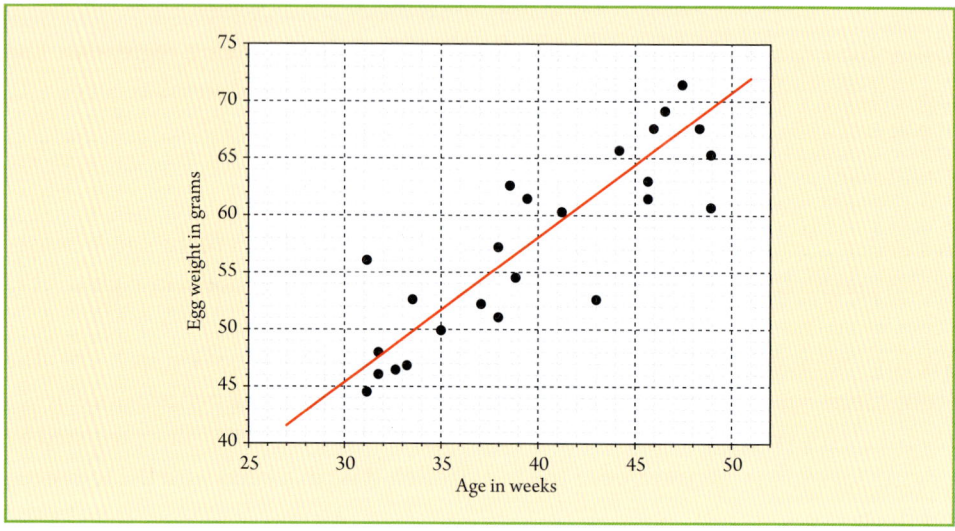

Figure 9.18. Plot of the test dataset and the regression line (model).

Another measure to quantify the performance of the model is R-squared (R^2). R^2 is defined as the portion of the total variability in the dataset explained by the model. For simple linear regression, the R^2 has a value between 0 and 1. A value of R^2 close to 0 means that the model has a low performance on the test dataset. This could mean that the relationship between dependent and independent variable is not linear or that more independent variables have to be considered. A higher value of R^2 means a better model (more variability in the test dataset is explained by the model).

For our test dataset, we find a R^2 of 0.72 meaning that more than 70% of the variability is explained by our model. What value of R^2 is considered as a good model performance depends on the domain you are working in. But as a rule of thumb, an R^2 above 0.7 qualifies as a good fit.

Be aware that our model is only valid for the range of the ages in the training dataset (from 21 to 53 weeks). Using the model to make predictions outside this range leads to erroneous results (according to our model a hen of 0 weeks would produce an egg of 7.3 grams!).

9.3.5 K-means clustering

Classification and decision trees are both supervised methods to classify data. You need a training set of labelled samples to use these methods. If you have samples without labels and you want to group them into clusters with similar characteristics, you have to use a clustering method. A frequently used clustering method is k-means clustering where k stands for the number of clusters you want to identify in your dataset.

In k-means clustering, you have to select a distance metric to quantify the distance between observations. The selection of the distance metric has to be done carefully as the outcomes of clustering algorithms may depend on the selected distance metric. Frequently used distance metrics are Euclidean distance, Manhattan distance and Mahalanobis distance (see Section 9.3.2).

K-means clustering starts with a random selection of k distinct observations. These observations serve as the initial cluster centres. Then we take the first observations in our dataset and measure the distance to each of the cluster centres and assign the observation to the cluster that is closest by. This process is repeated until all observations in the dataset are assigned to a cluster. Then we calculate the mean for each cluster and repeat the step of assigning observations to a cluster. If the clustering doesn't change, we are finished.

We can quantify the quality of the clustering by calculating the within-cluster variability. The within-cluster variability is defined as the sum of the distances (d) from each observation (x_j) to the centre of the cluster (x_{mean}) the observation belongs to:

$$Within\ cluster\ variability = \sum_{j=1}^{n} d(x_j, x_{mean})$$

where n is the number of observations in the data set.

K-means clustering is not deterministic since the result depends on the initial random selection of your k distinct data points. We repeat the clustering every time with a different initial random selection. For every clustering, we calculate the within-cluster variability and we pick the clustering with the lowest within-cluster variability as our best clustering.

In clustering, the number of clusters is not known beforehand. If you cannot decide on the number of clusters based on your domain knowledge, you have to find the optimal number of clusters from your dataset by running the clustering with an increasing number of clusters. You will notice that the within-cluster variability reduces when the number of clusters increases. To find the optimal number of clusters, you plot the within-cluster variability as a function of the number of clusters. Such a plot is called an elbow plot (Figure 9.19). At a certain number of clusters adding more clusters does not yield a strong reduction in within-cluster variability. This is the optimal number of clusters. In Figure 9.19 you notice that the within-cluster variability does not significantly reduce when the number of clusters is larger than four.

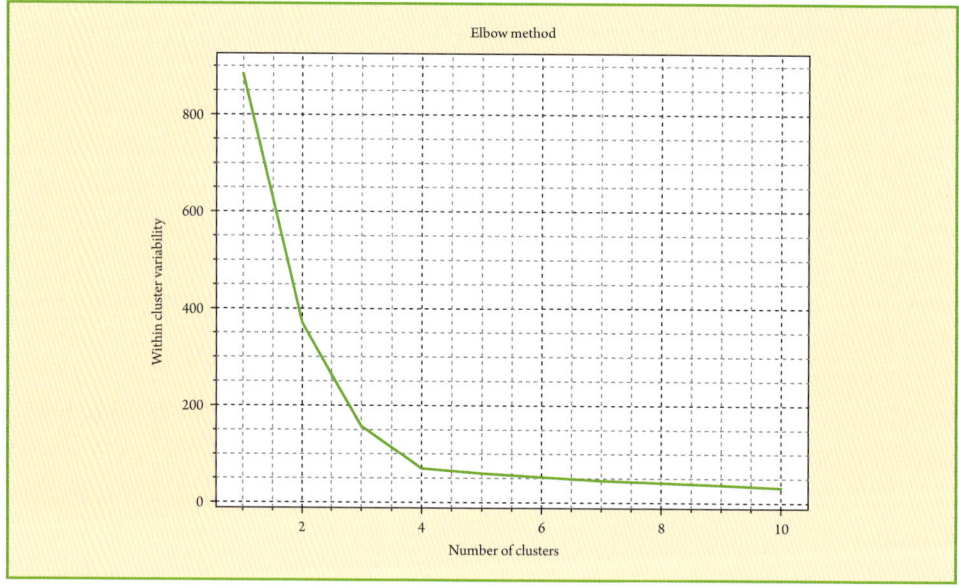

Figure 9.19. Elbow plot cluster analysis.

9.4 Tooling

The tooling used by data scientists can be classified into two groups:

- ▶ User-Interface (UI) based tools such as:
  - Excel (https://www.microsoft.com/en-us/microsoft-365/excel)
  - PowerBI (https://powerbi.microsoft.com/)
  - SPSS (https://www.ibm.com/products/spss-statistics)
  - Rapid Miner (https://rapidminer.com/)

- Scripting languages such as:
    - R (https://www.r-project.org/)
    - Python (https://www.python.org/)

The UI-based tools are in general easier to learn and more user-friendly than scripting languages. Scripting languages have the advantage that they can easily be extended with packages to add additional functionality and that they can be used to automate data science pipelines.

Both R and Python are free and open source and are supported by large user and developer communities. Both can easily be integrated with external tools such as Hadoop, a framework for distributed processing of large datasets (https://hadoop.apache.org/). For R there is RStudio as an integrated development environment (IDE). For Python, there are multiple IDE's to choose from, for example, PyCharm, Spider or Jupyter.

Python is a mature programming language with a clean and readable syntax supporting different programming paradigms, such as procedural and object-oriented programming. R is a programming language initially developed for statistical computing. It is said to have a more complex syntax than Python and therefore a steeper learning curve. Both languages offer a wide range of packages for data handling, statistical analysis and visualisation.

The classification, regression and clustering examples from Section 9.3 have been done with Python. The following packages were used:

- Pandas for importing data into tabular data structures from different sources, such as CSV, Excel and databases (https://pandas.pydata.org/).
- Matplotlib for creating histograms, boxplots and scatterplots (https://matplotlib.org/).
- NumPy for working with large, multi-dimensional arrays and matrices of data, along with a large collection of high-level mathematical functions to operate on these arrays (https://numpy.org/).
- SciPy for numerical integration, interpolation, optimisation, linear algebra, and statistics (https://www.scipy.org/scipylib/index.html).
- Scikit-learn for classification, regression and clustering algorithms. It interoperates with the Python numerical and scientific libraries NumPy and SciPy (https://scikit-learn.org/stable/index.html).

9.5 Future trends in data science applications

New techniques which are currently being used or developed in other industries than livestock might have great future implications for both farm and companion animals alike. Techniques such as deep learning which is considered a subset of machine learning, use neural networks that enable machines to make accurate decisions without the help of humans (Grossfeld, 2020). This neural network is a multi-layered system of artificial neurons with a structure similar to how humans would draw conclusions. This mimicking of the neural network of the brain leads to a process of learning that is far more capable than that of a standard machine learning model.

Also, the growing application of machine vision and with-it object recognition hold promise in livestock rearing (Nasirahmadi *et al.*, 2017). Automatic computer vision systems could aid farmers and researchers with problems such as the monitoring of animals for, e.g. visual scoring, animal weighing and other routine tasks which are both time-consuming and costly. A machine vision approach is a cost-effective, non-stressful, non-invasive, and more objective method for monitoring animal behaviours which can be adapted to different types of animals in a variety of situations, using the animals' natural features (e.g. shape, colour, movement). Not only does it provide the possibility to monitor animal behaviour and hence animal welfare in real-time, but it could also change our perspective on breeding animals. For example, if machine vision could monitor the behaviour of individual chicks in a large barn, we would be able to determine the genetic background of certain behaviours, e.g. feather picking, and eradicate these issues through selective breeding.

References

Chapman, P., Clinton, J., Kerber, R., Khabaza, T., Reinartz, T., Shearer, C., Wirth, R., 2000. CRISP-DM 1.0: Step-by-step data mining guide. SPSS, Chicago, IL, USA. Available at: https://tinyurl.com/xvnz4wn4

Clifton, C., 2019. Data mining | computer science. Available at: https://www.britannica.com/technology/data-mining

Gartner, 2020. Definition of big data. Available at: https://www.gartner.com/en/information-technology/glossary/big-data

Grossfeld, B., 2020. Deep learning vs machine learning: a simple way to understand the difference. Available at: https://www.zendesk.com/blog/machine-learning-and-deep-learning/

Hayashi, C., 1998. What is data science? Fundamental concepts and a heuristic example. In: Hayashi, C., Yajima, K., Bock, H.-H., Ohsumi, N., Tanaka, Y., Baba, Y. (eds.). Data science, classification, and related methods. Studies in classification, data analysis, and knowledge organization. Springer, Japan, pp. 40-51. https://doi.org/10.1007/978-4-431-65950-1

Leek, J., 2013. The key word in 'data science' is not data, it is science. Available at: https://simplystatistics.org/2013/12/12/the-key-word-in-data-science-is-not-data-it-is-science/

Mitchell, T.M., 1997. Machine learning. McGraw Hill, New York, NY, USA.

Nasirahmadi, A., Edwards, S.A., Sturm, B., 2017. Implementation of machine vision for detecting behaviour of cattle and pigs. Livestock Science 202: 25-38 http://dx.doi.org/10.1016/j.livsci.2017.05.014

Provost, F., Fawcett, T., 2013. Data science for business. O'Reilly Media Inc., Sebastopol, USA, 384 pp.

Visser, E.K., Van Wijk-Jansen, E.E.C., 2012. Diversity in horse enthusiasts with respect to horse welfare: an explorative study. Journal of Veterinary Behavior 7: 295-304.